Lecture Notes in Computer Science 14953

Founding Editor

Juris Hartmanis

Series Editor

Gerhard Goos, *Karlsruhe Institute of Technology, Karlsruhe, Germany*

Editorial Board Members

Elisa Bertino, *Purdue University, West Lafayette, USA*
Wen Gao, *Peking University, Beijing, China*
Bernhard Steffen ⓘ, *TU Dortmund University, Dortmund, Germany*
Moti Yung ⓘ, *Columbia University, New York, USA*

The series Lecture Notes in Computer Science (LNCS), including its subseries Lecture Notes in Artificial Intelligence (LNAI) and Lecture Notes in Bioinformatics (LNBI), has established itself as a medium for the publication of new developments in computer science and information technology research, teaching, and education.

LNCS enjoys close cooperation with the computer science R & D community, the series counts many renowned academics among its volume editors and paper authors, and collaborates with prestigious societies. Its mission is to serve this international community by providing an invaluable service, mainly focused on the publication of conference and workshop proceedings and postproceedings. LNCS commenced publication in 1973.

Kazuhiro Ogata · Narciso Martí-Oliet
Editors

Rewriting Logic and Its Applications

15th International Workshop, WRLA 2024
Luxembourg City, Luxembourg, April 6–7, 2024
Revised Selected Papers

Editors
Kazuhiro Ogata
Japan Advanced Institute of Science
and Technology
Nomi, Japan

Narciso Martí-Oliet
Universidad Complutense de Madrid
Madrid, Spain

ISSN 0302-9743   ISSN 1611-3349 (electronic)
Lecture Notes in Computer Science
ISBN 978-3-031-65940-9   ISBN 978-3-031-65941-6 (eBook)
https://doi.org/10.1007/978-3-031-65941-6

© The Editor(s) (if applicable) and The Author(s), under exclusive license to Springer Nature Switzerland AG 2024

This work is subject to copyright. All rights are solely and exclusively licensed by the Publisher, whether the whole or part of the material is concerned, specifically the rights of translation, reprinting, reuse of illustrations, recitation, broadcasting, reproduction on microfilms or in any other physical way, and transmission or information storage and retrieval, electronic adaptation, computer software, or by similar or dissimilar methodology now known or hereafter developed.
The use of general descriptive names, registered names, trademarks, service marks, etc. in this publication does not imply, even in the absence of a specific statement, that such names are exempt from the relevant protective laws and regulations and therefore free for general use.
The publisher, the authors and the editors are safe to assume that the advice and information in this book are believed to be true and accurate at the date of publication. Neither the publisher nor the authors or the editors give a warranty, expressed or implied, with respect to the material contained herein or for any errors or omissions that may have been made. The publisher remains neutral with regard to jurisdictional claims in published maps and institutional affiliations.

This Springer imprint is published by the registered company Springer Nature Switzerland AG
The registered company address is: Gewerbestrasse 11, 6330 Cham, Switzerland

If disposing of this product, please recycle the paper.

Preface

This volume contains the post-proceedings of the 15th International Workshop on Rewriting Logic and Its Applications (WRLA 2024).

Rewriting logic is a natural model of computation and an expressive semantic framework for concurrency, parallelism, communication, and interaction. It can be used to specify and verify a wide range of systems with their desired properties, and define domain-specific languages in various application fields. It also has good properties as a metalogical framework for representing logic. Over the years, several languages based on rewriting logic have been designed and implemented. The aim of the workshop is to bring together researchers with a common interest in rewriting logic and its applications, and to give them the opportunity to present their recent work, discuss future research directions, and exchange ideas. The previous meetings were held in Asilomar (USA) 1996, Pont-à-Mousson (France) 1998, Kanazawa (Japan) 2000, Pisa (Italy) 2002, Barcelona (Spain) 2004, Vienna (Austria) 2006, Budapest (Hungary) 2008, Paphos (Cyprus) 2010, Tallinn (Estonia) 2012, Grenoble (France) 2014, Eindhoven (The Netherlands) 2016, Thessaloniki (Greece) 2018, online 2020 (during COVID-19), and Munich (Germany) 2022.

WRLA 2024 was held during April 6–7, 2024, in Luxembourg City, Luxembourg, as a satellite event of the European Joint Conferences on Theory and Practice of Software (ETAPS 2024). We received 16 submissions; two were withdrawn due to being out of scope, and one was withdrawn due to being outside the submission categories; each of the remaining (13 submissions) was single-blindly reviewed by at least three program committee members. After an extensive discussion, the program committee decided to accept 11 papers for inclusion in this LNCS volume. The 11 papers are categorized as eight regular papers, two tool papers, and one education paper. This LNCS volume also includes the abstracts of the five invited tutorials delivered at WRLA 2024.

We sincerely thank all the authors of the papers submitted to the workshop, and the invited speakers for kindly agreeing to contribute to WRLA 2024. We are grateful to the members of the program committee and the additional reviewers for their careful work in the review process. We also thank the members of the WRLA steering committee for their valuable suggestions. Finally, we express our gratitude to all members of the local organization of ETAPS 2024, whose work made the workshop possible.

Nomi, Japan
Madrid, Spain
May 2024

Kazuhiro Ogata
Narciso Martí-Oliet

Organization

Program Committee

Erika Abraham	RWTH Aachen University, Germany
Kyungmin Bae	Pohang University of Science and Technology, South Korea
Canh Minh Do	Japan Advanced Institute of Science and Technology, Japan
Francisco Durán	University of Málaga, Spain
Santiago Escobar	Universitat Politècnica de València, Spain
Maribel Fernández	King's College London, UK
Nao Hirokawa	JAIST, Japan
Alexander Knapp	Universität Augsburg, Germany
Temur Kutsia	Johannes Kepler University Linz, Austria
Alberto Lluch-Lafuente	Technical University of Denmark, Denmark
Dorel Lucanu	Alexandru Ioan Cuza University, Romania
Salvador Lucas	Universitat Politècnica de València, Spain
Narciso Martí-Oliet	Universidad Complutense de Madrid, Spain
José Meseguer	University of Illinois at Urbana-Champaign, USA
Aart Middeldorp	University of Innsbruck, Austria
Masaki Nakamura	Toyama Prefectural University, Japan
Kazuhiro Ogata	JAIST, Japan
Peter Csaba Ölveczky	University of Oslo, Norway
Adrián Riesco	Universidad Complutense de Madrid, Spain
Christophe Ringeissen	INRIA, France
Camilo Rocha	Pontificia Universidad Javeriana Cali, Colombia
Traian-Florin Serbanuta	University of Bucharest, Romania
Carolyn Talcott	SRI International, USA

Publicity Chair

Duong Dinh Tran	JAIST, Japan

Additional Reviewers

Rubio, Rubén
Sapiña, Julia
Tran, Duong Dinh

Contents

Regular Papers

Verifying Invariants by Deductive Model Checking 3
  Kyungmin Bae, Santiago Escobar, Raúl López-Rueda, José Meseguer, and Julia Sapiña

Time-Bounded Resilience 22
  Tajana Ban Kirigin, Jesse Comer, Max Kanovich, Andre Scedrov, and Carolyn Talcott

Verifying Safe Memory Reclamation in Concurrent Programs
with CafeOBJ 45
  Duong Dinh Tran and Kazuhiro Ogata

Equivalence, and Property Internalization and Preservation for Equational
Programs 62
  José Meseguer

Equivalence Checking of Quantum Circuits Based on Dirac Notation
in Maude 84
  Canh Minh Do and Kazuhiro Ogata

Unified Opinion Dynamic Modeling as Concurrent Set Relations
in Rewriting Logic 104
  Carlos Olarte, Carlos Ramírez, Camilo Rocha, and Frank Valencia

Timed Strategies for Real-Time Rewrite Theories 124
  Carlos Olarte and Peter Csaba Ölveczky

Specifying Fairness Constraints and Model Checking with Non-intensional
Strategies 145
  Rubén Rubio, Narciso Martí-Oliet, Isabel Pita, and Alberto Verdejo

Tool Papers

The Hrewrite Library: A Term Rewriting Engine for Automatic Code
Assembly 165
  Michael Lienhardt

A Flexible Framework for Integrating Maude and SMT Solvers Using Python 179
   Geunyeol Yu and Kyungmin Bae

Education Papers

Teaching an Advanced Maude-Based Formal Methods Course in Oslo 195
   Peter Csaba Ölveczky

Author Index 209

Regular Papers

Verifying Invariants by Deductive Model Checking

Kyungmin Bae[1], Santiago Escobar[2], Raúl López-Rueda[2], José Meseguer[3(✉)], and Julia Sapiña[2]

[1] Pohang University of Science and Technology, Pohang, South Korea
kmbae@postech.ac.kr
[2] VRAIN, Universitat Politècnica de València, Valencia, Spain
{sescobar,rloprue,jsapina}@upv.es
[3] University of Illinois at Urbana-Champaign, Urbana, IL, USA
meseguer@illinois.edu

Abstract. We propose a new *deductive model checking* methodology where narrowing-based logical model checking of symbolic states specified as disjunctions of *constrained patterns* is synergistically combined with inductive theorem proving to discharge inductive verification conditions. An obvious synergy is to use an inductive theorem prover in *automated mode* as an *oracle* to help logical model checking reach a fixpoint. But this is not the only possible synergy. In this paper we focus instead on a new deductive model checking methodology to verify invariants—including inductive invariants—of infinite-state systems, where logical model checking automates large parts of the verification effort with the help of an inductive theorem prover as an *oracle*, and the remaining tasks are left to the inductive theorem prover used in interactive mode. We demonstrate this methodology by means of Maude examples using two tools working in tandem: the **DM-Check** symbolic model checker, and the **NuITP** inductive theorem prover.

S Escobar, J. Sapiña and R. López-Rueda have been partially supported by the grant PID2021-122830OB-C42 funded by MCIN/AEI/10.13039/501100011033 and ERDF A way of making Europe, and by the grant CIPROM/2022/6 funded by Generalitat Valenciana. S Escobar and K. Bae have been partially supported by the NATO Science for Peace and Security Programme project SymSafe (grant number G6133). J. Meseguer, S Escobar, J. Sapiña and R. López-Rueda have been partially supported by INCIBE's Chair funded by the EU-NextGenerationEU through the Spanish government's Plan de Recuperación, Transformación y Resiliencia. K. Bae has been partially supported by the National Research Foundation of Korea (NRF) grants funded by the Korean government (2021R1A5A1021944 and RS-2023-00251577).

© The Author(s), under exclusive license to Springer Nature Switzerland AG 2024
K. Ogata and N. Martí-Oliet (Eds.): WRLA 2024, LNCS 14953, pp. 3–21, 2024.
https://doi.org/10.1007/978-3-031-65941-6_1

1 Introduction

Amir Pnueli gave a fascinating invited talk entitled "Deduction is Forever" at FM'99 [24]. One of his main points was that, although explicit-state model checking was by then widely adopted in circuit design, deductive verification was essential for scalability; and a judicious combination of model checking and theorem proving was needed. This paper is all about synergistically combining logical model checking and inductive theorem proving into what we call *deductive model checking*: a combination where both formal methods cooperate so intimately that their differences almost evaporate. The key link relating both formal methods is *narrowing*, an automated deduction technique originating in resolution-based theorem proving [27]. Although originally introduced as an efficient *paramodulation* method to symbolically reason about equality, narrowing was later generalized in [16] as a formal method, not just in equational logic, but in rewriting logic [17], to symbolically perform *reachability analysis* of concurrent systems, i.e., as a logical model checking method to verify infinite-state systems. These narrowing-based logical model checking techniques have been further advanced by (see [1,9,11]): (i) symbolic state space reduction techniques; (ii) an extension from reachability analysis to LTL model checking; (iii) special techniques for cryptographic protocol verification; and (iv) symbolic model checkers. Maude itself also supports narrowing-based infinite-state model checking [7]. However, in all the just-mentioned work the system specifications (rewrite theories) that can be model checked, and the state predicates that can be used to specify properties are quite restricted: only unconditional rewrite theories $\mathcal{R} = (\Sigma, E \cup B, R)$ whose equations $E \cup B$ have the finite variant property (FVP) [12] can be model checked; and the only state predicates allowed are constructor terms $u(x_1, \ldots, x_n)$, called *constructor patterns*, denoting the set of concrete states obtained from $u(x_1, \ldots, x_n)$ by instantiating the variables $x_1, \ldots, x_n$ by ground terms.

The situation has significantly changed for the better thanks to [20]: (i) the rewrite theories that can be analyzed can have conditional rules and have virtually no restrictions; (ii) the state predicates are now generalized to *constrained constructor patterns* of the form $u(x_1, \ldots, x_n) \mid \varphi$, with φ a quantifier-free formula, describing the instances of $u(x_1, \ldots, x_n)$ that satisfy φ; furthermore, any finite unions and/or intersections of such sets can always be described by a *finite disjunction* of constrained constructor patterns, an expressive symbolic language to specify pre- and post-conditions; (iii) narrowing is generalized to *constrained narrowing*, where the sets of states thus narrowed by the possibly conditional rules R of the rewrite theory $\mathcal{R} = (\Sigma, E \cup B, R)$ specifying the system's transition relation are specified as disjunctions of constrained patterns; and (iv) constrained narrowing computes the *predicate transformer* $R(\_)$ (see [20]) that maps a disjunction A of constrained patterns to a resulting such disjunction $R(A)$ denoting the states reachable from A in one R-transition step.

This greater generality and expressiveness brings with it a natural marriage between logical model checking and *inductive theorem proving*, which is needed for state space reduction by *folding*. The set of states R-*reachable* from a disjunction A of constrained constructor patterns is structured as a (usually infinite) *narrowing forest* built in a breadth first manner. Folding tries to transform this infinite forest into a finite graph (a fixpoint) by *subsuming* (and therefore deleting) a new symbolic state $u \mid \varphi$ generated at depth $k+1$ as a substitution instance of a previously generated symbolic state at depth $j \leq k+1$. For unconstrained patterns u and v, a B-matching algorithm (for B the structural axioms in $\mathcal{R} = (\Sigma, E \cup B, R)$) is used. For a constrained constructor pattern $u \mid \varphi$, checking whether it is subsumed by a previously generated one $v \mid \psi$ generally requires inductive theorem proving (see Sect. 2). In the deductive model checking approach that we propose, symbolic model checking and inductive theorem proving work *together* in two modes: (a) an *automatic* one, where a terminating strategy σ of a theorem prover is used as an *oracle* to check whether a constrained pattern $u \mid \varphi$ can be folded into another one $v \mid \psi$; and (b) an *interactive* mode, in which verification conditions that could not be automatically checked are handled by the inductive theorem prover. But inductive theorem proving is not just a supporting actor in the cast, with symbolic model checking as the protagonist. Their synergistic combination opens up new possibilities where the roles are changed.

This is nicely illustrated by the main focus of this paper, namely, the application of deductive model checking to the *verification of invariants*, including inductive ones. After some preliminaries in Sect. 2 we:

1. Present in Sect. 3 a *proof methodology* for verifying invariants where, denoting as A, B, C, etc., disjunctions of constrained patterns, we show how to prove by a *mixture of automatic and interactive methods* that: (i) a set A of initial constrained terms is contained in a conjectured invariant denoted by the set B of constrained terms; (ii) B is an *inductive*, i.e., transition-closed, invariant; (iii) C is an invariant because it contains a verified invariant B, which can be proved in two ways: (iii).1 *positively*, by proving such containment, and (iii).2 *negatively*, by showing that the states in the complement of C have no states in common with those denoted by B.
2. In Sect. 4 we explain and illustrate with an example how the invariant proof methodology presented in Sect. 3 is supported by the prototype constrained narrowing model checker **DM-Check**, publicly available[1], working in tandem with Maude's prototype **NuITP** inductive theorem prover [6].
3. In Sect. 5 a case study on the Alternating Bit Protocol (ABP) is presented.
4. In Sect. 6 we discuss related work and in Sect. 7 we present some conclusions.

2 Preliminaries

We assume familiarity with the notions of an order-sorted signature Σ on a poset of sorts $(S, \leq)$, an order-sorted Σ-algebra $\mathbb{A}$, and the term Σ-algebras $\mathbb{T}_\Sigma$ and $\mathbb{T}_\Sigma(X)$ for X an S-sorted set of variables. We also assume familiarity with the notions of: (i) Σ-homomorphism $h : \mathbb{A} \to \mathbb{B}$ between Σ-algebras $\mathbb{A}$ and $\mathbb{B}$,

---
[1] At http://safe-tools.dsic.upv.es/dmc.

so that Σ-algebras and Σ-homomorphisms form a category $\mathbf{OSAlg}_\Sigma$; (ii) order-sorted (i.e., sort-preserving) substitution θ, its domain $dom(\theta)$ and range $ran(\theta)$, and its application $t\theta$ to a term t; (iii) *preregular* order-sorted signature Σ, i.e., a signature such that each term t has a least sort, denoted $ls(t)$; (iv) the set $\widehat{S} = S/(\geq \cup \leq)^+$ of *connected components* of a poset $(S, \leq)$ viewed as a DAG; (v) for $\mathbb{A}$ a Σ-algebra, the set A_s of its elements of sort $s \in S$; (vi) the order-sorted equational deduction relation $E \vdash u = v$ and its associated E-equality relation $=_E$; (vi) *E-unifiers* of an equation $u = v$, i.e., substitutions θ such that $u\theta =_E v\theta$; and (vii) the satisfaction relation $\mathbb{A} \models \varphi$ of a first-order Σ-formula φ by a Σ-algebra $\mathbb{A}$. We furthermore assume that all signatures Σ have *non-empty sorts*, i.e., $T_{\Sigma,s} \neq \emptyset$ for each $s \in S$. $[A \to B]$ denotes the *S-sorted functions* from A to B. These notions are explained in detail in [14,18,19]. The material below is adapted from [19,20].

Convergent Theories and Sufficient Completeness. Given an order-sorted equational theory $\mathcal{E} = (\Sigma, E \cup B)$, where B is a collection of associativity and/or commutativity and/or identity axioms and Σ is B-preregular, we can associate to it a corresponding *rewrite theory* [17] $\boldsymbol{\mathcal{E}} = (\Sigma, B, \boldsymbol{E})$ by orienting the equations E as left-to-right rewrite rules. That is, each $(u = v) \in E$ is transformed into a rewrite rule $u \to v$. For simplicity, we recall here the case of unconditional equations. Since in this work we will consider *conditional* theories $\boldsymbol{\mathcal{E}}$, we refer to [15] for full details on the general definition of convergent theory. The main purpose of the rewrite theory $\boldsymbol{\mathcal{E}}$ is to reduce the complex bidirectional reasoning with equations to the much simpler unidirectional reasoning with rules under suitable assumptions. We assume familiarity with the notion of subterm $t|_p$ of t at a term position p and of term replacement $t[w]_p$ of $t|_p$ by w at position p (see, e.g., [5]). The rewrite relation $t \to_{\boldsymbol{E},B} t'$ holds iff there is a subterm $t|_p$ of t, a rule $(u \to v) \in \boldsymbol{E}$ and a substitution θ such that $u\theta =_B t|_p$, and $t' = t[v\theta]_p$. We denote by $\to^*_{\boldsymbol{E},B}$ the reflexive-transitive closure of $\to_{\boldsymbol{E},B}$. For $\boldsymbol{\mathcal{E}}$ unconditional, the convergence requirements are as follows (see [15] for $\boldsymbol{\mathcal{E}}$ conditional): (i) $vars(v) \subseteq vars(u)$; (ii) *sort-decreasingness*: for each substitution θ, $ls(u\theta) \geq ls(v\theta)$; (iii) *strict B-coherence*: if $t_1 \to_{\boldsymbol{E},B} t'_1$ and $t_1 =_B t_2$ then there exists $t_2 \to_{\boldsymbol{E},B} t'_2$ with $t'_1 =_B t'_2$; (iv) *confluence* (resp. *ground confluence*) modulo B: for each term t (resp. ground term t) if $t \to^*_{\boldsymbol{E},B} v_1$ and $t \to^*_{\boldsymbol{E},B} v_2$, then there exist rewrite sequences $v_1 \to^*_{\boldsymbol{E},B} w_1$ and $v_2 \to^*_{\boldsymbol{E},B} w_2$ such that $w_1 =_B w_2$; (v) *termination*: the relation $\to_{\boldsymbol{E},B}$ is well-founded (for $\boldsymbol{E}$ conditional, we require *operational termination* [15]). If $\boldsymbol{\mathcal{E}}$ satisfies conditions (i)–(v) (resp. the same, but (iv) weakened to ground confluence modulo B), then it is called *convergent* (resp. *ground convergent*). The key point is that then, given a term (resp. ground term) t, all terminating rewrite sequences $t \to^*_{\boldsymbol{E},B} w$ end in a term w, denoted $t!_{\boldsymbol{\mathcal{E}}}$, that is unique up to B-equality, and its called t's *canonical form*. Ground convergence implies three major results: (1) for any ground terms t, t' we have $t =_{E \cup B} t'$ iff $t!_{\boldsymbol{\mathcal{E}}} =_B t'!_{\boldsymbol{\mathcal{E}}}$, (2) the B-equivalence classes of canonical forms are the elements of the *canonical term algebra* $C_{\Sigma/E,B}$, where for each $f : s_1 \ldots s_n \to s$ in Σ and B-equivalence classes of canonical terms $[t_1], \ldots, [t_n]$ with $ls(t_i) \leq s_i$ the operation

$f_{C_{\Sigma/E,B}}$ is defined by the identity: $f_{C_{\Sigma/E,B}}([t_1]\ldots[t_n]) = [f(t_1\ldots t_n)!_{\mathcal{E}}]$, and (3) we have an isomorphism $T_{\mathcal{E}} \cong C_{\Sigma/E,B}$.

A ground convergent rewrite theory $\mathcal{E} = (\Sigma, B, E)$ is called *sufficiently complete* with respect to a subsignature Ω, whose operators are then called *constructors*, iff for each ground Σ-term t, $t!_{\mathcal{E}} \in T_{\Omega}$. Furthermore, for $\mathcal{E} = (\Sigma, B, E)$ sufficiently complete w.r.t. Ω, a ground convergent rewrite subtheory $(\Omega, B_{\Omega}, E_{\Omega}) \subseteq (\Sigma, B, E)$ is called a *constructor subspecification* iff $T_{\mathcal{E}}|_{\Omega} \cong T_{\Omega/E_{\Omega} \cup B_{\Omega}}$. If $E_{\Omega} = \emptyset$, then Ω is called a signature of *free constructors modulo axioms* B_{Ω}. Note that $\mathcal{E} = (\Sigma, B, E)$ is sufficiently complete with respect to Ω iff each ground Σ-term $t \in T_{\Sigma} \setminus T_{\Omega}$ is E, B-reducible, i.e., $t \rightarrow_{E,B} t'$ holds for some $t' \in T_{\Sigma}$.

Specifying Sets of States with Constrained Constructor Patterns. We summarize here the main ideas and results in [20] on constrained constructor patterns. Let $(\Sigma, E \cup B)$ be an equational theory having a ground convergent decomposition (Σ, B, E), a constructor subtheory $(\Omega, B_{\Omega}, \emptyset) \subseteq (\Sigma, B, E)$, and a sort *State* at the top of one of its connected components whose data elements denote states of a rewrite theory having (Σ, B, E) as its equational subtheory. Constrained constructor patterns are an expressive language to specify sets of states in the data type $\mathbb{T}_{\Omega/B_{\Omega}}$ of constructors of (Σ, B, E). The language defines a distributive lattice (see [20]) generated by constrained constructor patterns of the form $u \mid \varphi$, where $u \in T_{\Omega, State}(X)$, with X_s countable for each sort s in Σ, and φ a conjunction of Σ-equalities. The language is then the closure of such patterns under the $\vee$ and $\wedge$ lattice operations. We shall use capital letters $A, B, C, \ldots$ to describe expressions in such a distributive lattice. The *semantics* of any such $u \mid \varphi$ is the set of states:

$$[\![u \mid \varphi]\!] = \{[v] \in \mathbb{T}_{\Omega/B_{\Omega}, State} \mid \exists \rho \in [X \rightarrow T_{\Omega}] \text{ s.t. } [v] = [u\rho] \wedge E \cup B \models \varphi\rho\}.$$

Such a mapping extends to equivalence classes of expressions $A, B, C, \ldots$ and defines a homomorphism of distributive lattices so that:

$$[\![A \vee B]\!] = [\![A]\!] \cup [\![B]\!] \quad \text{and} \quad [\![A \wedge B]\!] = [\![A]\!] \cap [\![B]\!].$$

Every expression A has a semantically equivalent expression involving only basic constrained patterns closed only under $\vee$, because of the semantic identities:

$$[\![u \mid \varphi \wedge v \mid \psi]\!] = [\![u \mid \varphi]\!] \cap [\![v \mid \psi]\!] = \bigcup_{\alpha \in DUnif_{B_{\Omega}}(u=v)} [\![(u \mid \varphi \wedge \psi)\alpha]\!]$$

where $DUnif_{B_{\Omega}}(u = v)$ denotes a complete set of *disjoint*[2] B_{Ω}-unifiers of the equation $u = v$. Set theoretic inclusion $[\![u \mid \varphi]\!] \subseteq [\![v \mid \psi]\!]$ has an attractive *sufficient condition* called B_{Ω}-*subsumption* and denoted $u \mid \varphi \sqsubseteq_{B_{\Omega}} v \mid \psi$, which by definition holds iff there exists a B_{Ω}-matching substitution α such that: (i) $u =_{B_{\Omega}} (v\alpha)$, and (ii) $\mathbb{T}_{\Sigma/E \cup B} \models \varphi \Rightarrow (\psi\alpha)$. As illustrated by an example in Sect. 3, subsumption is not a necessary condition for set containment.

---

[2] A disjoint E-unifier of $u = v$ is an E unifier of $u' = v'$, with u' (resp. v') a variable renaming of u (resp. v) such that u' and v' share no variables.

The language of constrained patterns is quite expressive, since it provides a constructive version of set theory when we view $[\![u \mid \varphi]\!]$ as the set $\{u(x_1,\ldots,x_n) \in T_{\Omega/B_\Omega,State} \mid \varphi(x_1,\ldots,x_n)\}$. Nothing prevents $(\Sigma, B, \boldsymbol{E})$ from specifying various computable *auxiliary functions* to make such a language even more expressive.

Logical Model Checking with Constrained Narrowing. Maude provides narrowing-based model checking of concurrent systems specified as *topmost*[3] rewrite theories $\mathcal{R} = (\Sigma, E \cup B, R)$ [7], but only under quite restrictive requirements, namely: (a) its equational subtheory $(\Sigma, E \cup B)$ must enjoy the finite variant property (FVP) [12]; and (b) its rules R must be *unconditional*. This means that only *unconditional patterns* are narrowed, i.e., in the above notation patterns the form $u_1 \mid \top$. This severely limits both the systems and the properties that can be specified and verified. The situation has drastically changed after [20] for two reasons: (i) now fully general *admissible*[4] topmost rewrite theories $\mathcal{R} = (\Sigma, E \cup B, R)$ with FVP constructor subtheory $(\Omega, E_\Omega \cup B_\Omega)$ can be symbolically model checked; and (ii) the language of state predicates can be the expressive language of constrained constructor patterns.

Achieving logical, narrowing-based model checking in practice with the full generality of (i)–(ii) cannot be done using $\mathcal{R}$ directly. This is because, given a rule $l \to r$ *if* φ in R, it is safe to assume that l is an Ω-term, but r can often be a general Σ term, for which $E \cup B$-unification is both infinitary and undecidable. However, the generality of (i)–(ii) can be achieved *indirectly* by using instead the semantically equivalent theory $\mathcal{R}_{l,r}^\Omega$ defined in [20]. Therefore, from now on we shall assume that $\mathcal{R} = (\Sigma, E \cup B, R)$ is already of the form $\mathcal{R}_{l,r}^\Omega$. Furthermore, because it affords a simpler exposition and a more efficient implementation we will further assume that in $(\Omega, E_\Omega \cup B_\Omega)$, $E_\Omega = \emptyset$, i.e., the constructors are *free* modulo B_Ω. Under such assumptions, the initial transition system defined by $\mathcal{R}$ [20] has the form $\mathbb{C}_\mathcal{R} = (\mathbb{C}_{\Sigma/E,B}, \to_\mathcal{R})$, with[5] $\mathbb{C}_{\Sigma/E,B}|_\Omega = \mathbb{T}_{\Omega/B_\Omega}$.

For a topmost $\mathcal{R}$ satisfying these assumptions *constrained narrowing* [20] is a labeled relation between constrained constructor terms $u \mid \varphi \overset{\alpha}{\leadsto}_{R,B_\Omega} v \mid \psi$ that holds iff there exists a rule $l \to r$ *if* ϕ in R and a disjoint B_Ω-unifier α of the equation $u = l$ such that $v \mid \psi = (r \mid \varphi \wedge \phi)\alpha$. Its reflexive transitive closure $u \mid \varphi \overset{\alpha\,*}{\leadsto}_{R,B_\Omega} v \mid \psi$ holds if there is a number k and a sequence $u \mid \varphi \overset{\alpha_1}{\leadsto}_{R,B_\Omega} u_1 \mid \varphi_1 \ldots u_{k-1} \mid \varphi_{k-1} \overset{\alpha_k}{\leadsto}_{R,B_\Omega} u_k \mid \varphi_k$ such that $\alpha = \alpha_1 \ldots \alpha_k$, and $v \mid \psi = u_k \mid \varphi_k$. For $k = 0$ we just have $u \mid \varphi \overset{id}{\leadsto}_{R,B_\Omega} u \mid \varphi$, with *id* the identity substitution. For

---

[3] A rewrite theory $\mathcal{R}$ is *topmost* iff it has a sort *State* such that: (i) if $f(t_1,\ldots,t_n)$ has sort *State*, none of the t_i, $1 \leq i \leq n$ has sort *State*, and (ii) for any rewrite rule $l \to r$ *if* φ in $\mathcal{R}$, l and r have sort *State*. Many common rewrite theories such as, e.g., actor systems, can be transformed into semantically equivalent topmost ones.

[4] $\mathcal{R} = (\Sigma, E \cup B, R)$ is *admissible* if $(\Sigma, E \cup B)$ is ground confluent and the rules R are *ground coherent* w.r.t. the oriented equations $\boldsymbol{E}$ modulo B, i.e., for each $t \in T_{\Sigma,State}$ if $t \to_{R,B} t'$, then there exists a u' such that $t!_{E,B} \to_{R,B} u'$ and $t'!_{E,B} =_B u'!_{E,B}$.

[5] Given a Σ-algebra $\mathbb{A}$ and a subsignature Σ' with same poset of sorts, the *reduct* $\mathbb{A}|_{\Sigma'}$ is the Σ'-algebra with same sorts as $\mathbb{A}$ and same operations as $\mathbb{A}$ for each $f \in \Sigma'$.

technical reasons we assume that the variables of each $u_{i+1} \mid \varphi_{i+1}$ are always *fresh* w.r.t. those of any previous $u_j \mid \varphi_j$, $1 \leq j \leq 0$, with $u_0 \mid \varphi_0 =_{def} u \mid \varphi$. The key result in [20] ensuring the completeness of symbolic reachability analysis by constrained narrowing is the following theorem:

Theorem 1. ([20]) *Let $\mathcal{R}$ be topmost with equational theory $(\Sigma, E \cup B)$ and satisfy the requirements above, and let $\bigvee_{i \in I} u_i \mid \varphi_i$ and $\bigvee_{j \in J} v_j \mid \psi_j$ be finite disjunctions of constrained constructor patterns for $(\Sigma, E \cup B)$. Then, there exists $[w] \in [\![\bigvee_{i \in I} u_i \mid \varphi_i]\!]$ and $[w'] \in [\![\bigvee_{j \in J} v_j \mid \psi_j]\!]$ such that $[w] \to_{\mathcal{R}}^k [w']$ iff there are $i \in I$ and $j \in J$, a narrowing sequence $u_i \mid \varphi_i \overset{\alpha\ k}{\leadsto}_{R,B_\Omega} v \mid \psi$, a disjoint B_Ω-unifier β of the equation $v = v_j$ and a ground substitution ρ such that $E \cup B \vdash (\psi \wedge \psi_j)\beta\rho$.*

Constrained narrowing is a form of *deductive model checking*, where symbolic model checking and inductive theorem proving are synergistically combined, for three reasons: (1) The existence of a ground substitution ρ such that $E \cup B \vdash (\psi \wedge \psi_j)\beta\rho$ is another way to say that $\mathbb{T}_{\Sigma/E \cup B} \models \exists (\psi \wedge \psi_j)\beta$, where $\exists (\psi \wedge \psi_j)\beta$ denotes the existential closure of $(\psi \wedge \psi_j)\beta$; an *inductive* satisfiability property. (2) Constrained narrowing search from $\bigvee_{i \in I} u_i \mid \varphi_i$ to find an intersection with $\bigvee_{j \in J} v_j \mid \psi_j$ will usually generate an infinite *narrowing forest* that symbolically describes the set of all states reachable from $[\![\bigvee_{i \in I} u_i \mid \varphi_i]\!]$. However, if using the subsumption relation $\sqsubseteq_{B_\Omega}$ we can *fold* such an infinite narrowing tree into a *finite* narrowing graph by B_Ω-subsuming if possible any constrained pattern found at depth $k+1$ by some other constrained pattern found at depth $j \leq k$, the situation becomes much better, since only a finite number of intersections with a goal state $\bigvee_{j \in J} v_j \mid \psi_j$ need to be examined in finite time. But inductive theorem proving is of the essence for the subsumption relation $u \mid \varphi \sqsubseteq_{B_\Omega} v \mid \psi$ to hold, since for some matching substitution α we need to prove that $\mathbb{T}_{\Sigma/E \cup B} \models \varphi \Rightarrow (\psi\alpha)$, an inductive theorem. (3) Inductive theorem proving does not play second fiddle to logical model checking: it is an equal partner in system verification, i.e., new verification methods that combine the powers of logical model checking and inductive theorem proving and go beyond logical model checking come out of the merging of both formal methods. This is nicely illustrated by the proof methodology for verifying invariants proposed in Sect. 3 and demonstrated in Sects. 4 and 5.

3   A Proof Methodology for Verifying Invariants

As discussed in Sect. 2, without loss of generality we may assume that the admissible topmost rewrite theory $\mathcal{R} = (\Sigma, E \cup B, R)$ specifying our system of interest, and having a topmost sort *State*, is the result of transforming a general topmost rewrite theory $\mathcal{R}_0$ into a semantically equivalent one $\mathcal{R} = \mathcal{R}_{0_{l,r}}^\Omega$ such that its rewrite rules have the form: $u \to v$ *if* ϕ, with u and v terms in a constructor subsignature $\Omega \subseteq \Sigma$, which we assume free modulo the axioms B_Ω. An *invariant Q from* a set I of initial states is, by definition, a subset $Q \subseteq C_{\Sigma/E,B,State}$ such that: (i) $I \subseteq Q$, and (ii) $Reach_{\mathcal{R}}(I) \subseteq Q$, where

$Reach_\mathcal{R}(I) =_{def} \{[v] \in C_{\Sigma/E,B,State} \mid \exists [u] \in I \text{ s.t. } [u] \to^*_\mathcal{R} [v]\}$. Such a Q is called an *inductive invariant* from I if, in addition, (iii) Q is *transition-closed*, i.e., if $[u] \in Q$ and $[u] \to_\mathcal{R} [v]$, then $[v] \in Q$. Note that $Reach_\mathcal{R}(I)$ is the *smallest* inductive invariant from I.

Our invariant verification methodology for such a theory $\mathcal{R}$ combines constrained narrowing and inductive theorem proving to prove invariants as follows:

1. Specification of Initial States and of Invariants. Both the (typically) infinite) set of initial states, as well the set of states of a conjectured invariant Q (resp. of its complement Q^c) are symbolically specified as disjunctions of constrained constructor patterns of the form: $\bigvee_{i \in I} u_i \mid \varphi_i$.

2. Subsuming Initial States by a Conjectured Inductive Invariant. $\bigvee_{j \in J} v_j \mid \psi_j$ will be a (conjectured) inductive invariant *from* initial constrained patterns $\bigvee_{i \in I} u_i \mid \varphi_i$ if we prove the containment $\bigcup_{i \in I} [\![u_i \mid \varphi_i]\!] \subseteq \bigcup_{j \in J} [\![v_j \mid \psi_j]\!]$. Such a containment can be proved in two ways:

2.1. Automatically. Since: (i) the existence for each $i \in I$ of a $j \in J$ such that $[\![u_i \mid \varphi_i]\!] \subseteq [\![v_j \mid \psi_j]\!]$ is a *sufficient condition* for the above containment, (ii) the subsumption $u_i \mid \varphi_i \sqsubseteq_{B_\Omega} v_j \mid \psi_j$, i.e., the existence of a B-matching substitution α such that $u_i =_B (v_j \alpha)$ and $\mathbb{C}_{\Sigma/E,B} \models \varphi_i \Rightarrow (\psi_j \alpha)$ is a *sufficient condition* for the containment in (i), and (iii) a fixed, *terminating proof strategy* σ may succeed in proving that $\varphi_i \Rightarrow (\psi_j \alpha)$ is an inductive theorem, we may use (i), (ii) and σ to *automatically* prove that, for at least a *subset* $I_0 \subseteq I$, we have $\bigcup_{i \in I_0} [\![u_i \mid \varphi_i]\!] \subseteq \bigcup_{j \in J} [\![v_j \mid \psi_j]\!]$.

2.2. Deductively. Apart of proving the containment for I_0, we can use an *inductive theorem prover* to prove the remaining containment $\bigcup_{i \in I \setminus I_0} [\![u_i \mid \varphi_i]\!] \subseteq \bigcup_{j \in J} [\![v_j \mid \psi_j]\!]$. Two proof methods are possible: (i) For some $i \in I \setminus I_0$ the previous automatic method may have found a $j \in J$ such that $u_i =_B (v_j \alpha)$ but the automatic proof strategy σ could not prove that $\mathbb{C}_{\Sigma/E,B} \models \varphi_i \Rightarrow (\psi_j \alpha)$; if the user conjectures that such an implication is valid in $\mathbb{C}_{\Sigma/E,B}$, then an inductive theorem prover can be used to prove the containment $[\![u_i \mid \varphi_i]\!] \subseteq [\![v_j \mid \psi_j]\!]$. Method (i) may increase I_0 to a superset $I_1 \supseteq I_0$. We can then use the following Method (ii) to prove the remaining containment $\bigcup_{i \in I \setminus I_1} [\![u_i \mid \varphi_i]\!] \subseteq \bigcup_{j \in J} [\![v_j \mid \psi_j]\!]$ as follows: (ii).1 We add to Σ a signature Π containing a new sort *Pred* of predicates with a single constant tt of sort *Pred* and a predicate symbol, say, $p : State \to Pred$, (ii).2 we characterize the set $\bigcup_{j \in J} [\![v_j \mid \psi_j]\!]$ by the equations $E_p = \{p(v_j) = tt \text{ if } \psi_j\}_{j \in J}$, and (ii).3 we use an inductive theorem prover to prove the conjectures $\{\varphi_i \Rightarrow p(u_i) = tt\}_{i \in I \setminus I_1}$ in the initial algebra $\mathbb{C}_{\Sigma \cup \Pi / E \cup E_p, B}$.

As an example, let the elements of sort *State* be triples $\langle n, m, k \rangle$ of naturals in Peano notation, and suppose that we want to prove the containment $[\![\langle 0, y', s(z') \rangle \mid y' > s(0) = true]\!] \subseteq [\![\langle x, s(y), z \rangle \mid \top]\!]$. Predicate p is defined by the equation $p(\langle x, s(y), z \rangle) = tt$. Need to prove $y' > s(0) = true \Rightarrow p(\langle 0, y', s(z') \rangle) = tt$. Reasoning by cases $y' = 0$ or $y' = s(y'')$ gives us subgoals (1) $0 > s(0) = true \Rightarrow p(\langle 0, 0, s(z')\rangle) = tt$ and (2) $s(y'') > s(0) = true \Rightarrow p(\langle 0, s(y''), s(z')\rangle) = tt$. The first subgoal (1) is true due to its unsatisfiable condition; the second subgoal (2) is true because $p(\langle 0, s(y''), s(z')\rangle)$ reduces to tt.

3. Verifying an Inductive Invariant. If we have proved that $\bigvee_{j \in J} v_j \mid \psi_j$ is a conjectured inductive invariant *from* initial constrained patterns $\bigvee_{i \in I} u_i \mid \varphi_i$ by Method 2, then $\bigvee_{j \in J} v_j \mid \psi_j$ will actually be an *inductive invariant* from $\bigvee_{i \in I} u_i \mid \varphi_i$ iff the set $\bigcup_{j \in J} [\![v_j \mid \psi_j]\!]$ is *transition-closed*. By Theorem 1 (with k=1) such a set will be transition-closed iff for each $j \in J$, rule $l \to r$ if ϕ in R, and disjoint B_Ω-unifier $\alpha \in DUnif_B(l = v_j)$ we have $[\![(r \mid \psi_j \wedge \phi)\alpha]\!] \subseteq \bigcup_{j \in J} [\![v_j \mid \psi_j]\!]$. The method to prove such a containment is entirely similar to Method 2, except for one additional case: it may happen that for some $(r \mid \psi_j \wedge \phi)\alpha$ the constraint $(\psi_j \wedge \phi)\alpha$ is *unsatisfiable*, so that $[\![(r \mid \psi_j \wedge \phi)\alpha]\!] = \emptyset$, but the theorem proving strategy σ was not able to prove this fact. In such additional case, we use inductive theorem proving to prove that $\mathbb{C}_{\Sigma/E,B} \models \neg(\psi_j \wedge \phi)\alpha$.

4. Verifying other Invariants Positively. Suppose that we have already proved that $\bigvee_{j \in J} v_j \mid \psi_j$ is an inductive invariant from the initial constrained patterns $\bigvee_{i \in I} u_i \mid \varphi_i$, and we wish to prove that the set denoted by $\bigvee_{k \in K} w_k \mid \gamma_k$ is an invariant from $\bigvee_{i \in I} u_i \mid \varphi_i$. A sufficient condition for this to hold is proving the containment $\bigcup_{j \in J} [\![v_j \mid \psi_j]\!] \subseteq \bigcup_{k \in K} [\![w_k \mid \gamma_k]\!]$. The method to prove this containment is exactly Method 2. Of course, containment is only a sufficient condition. But in practice the cases of an invariant such that $\bigcup_{j \in J} [\![v_j \mid \psi_j]\!] \not\subseteq \bigcup_{k \in K} [\![w_k \mid \gamma_k]\!]$ should be rare, because often the set $\bigcup_{j \in J} [\![v_j \mid \psi_j]\!]$ coincides with $Reach_\mathcal{R}(\bigcup_{i \in I} [\![u_i \mid \varphi_i]\!])$, which forces the above containment.

5. Verifying other Invariants Negatively. Note that an invariant Q will contain an inductive invariant Q_0 iff $Q_0 \cap Q^c = \emptyset$. Furthermore, it is often simpler[6] to characterize Q^c by a disjunction of constrained patterns than to do so for Q. Suppose that $\bigvee_{j \in J} v_j \mid \psi_j$ is an inductive invariant from initial states $\bigvee_{i \in I} u_i \mid \varphi_i$ and that Q^c can be specified by a disjunction of constrained patterns $\bigvee_{k \in K} w_k \mid \gamma_k$. Then, a sufficient condition for Q to be an invariant from $\bigvee_{i \in I} u_i \mid \varphi_i$ is that $\bigcup_{j \in J} [\![v_j \mid \psi_j]\!] \cap \bigcup_{k \in K} [\![w_k \mid \gamma_k]\!] = \emptyset$. For the exact same reasons as in Method 4, though possible, it is unlikely that Q is an invariant from $\bigvee_{i \in I} u_i \mid \varphi_i$ and $\bigcup_{j \in J} [\![v_j \mid \psi_j]\!] \cap \bigcup_{k \in K} [\![w_k \mid \gamma_k]\!] \neq \emptyset$. The above intersection can be *computed symbolically*, namely, as the set of states denoted by $\bigvee_{j \in J, k \in K, \alpha \in DUnif_B(v_j = w_k)} (v_j \mid \psi_j \wedge \gamma_k)\alpha)$. If no such disjoint unifiers α exist, the containment is proved. Otherwise, we are left with a disjunction of constrained terms of the form $(v_j \mid \psi_j \wedge \gamma_k)\alpha$ for some $j \in J$, $k \in K$ and $\alpha \in DUnif_B(v_j = w_k)$ and the intersection will be empty if for each of them we can prove the inductive theorem $\mathbb{C}_{\Sigma/E,B} \models \neg(\psi_j \wedge \gamma_k)\alpha$. Note that if we have proved in this, negative way that Q is an invariant from $\bigvee_{i \in I} u_i \mid \varphi_i$, we have actually proved $2^{|K|} - 1$ invariants in one blow, namely, that the complement

---

[6] One case where this does not happen is deadlock-freedom, where a proof by the positive method is easier: see Sect. 4 for an example.

of the set of states denoted by $\bigvee_{k \in K_0} w_k \mid \gamma_k$ for any non-empty $K_0 \subseteq K$ is an invariant from $\bigvee_{i \in I} u_i \mid \varphi_i$.

4 Verifying Invariants with DM-Check and NuITP

DM-Check is a deductive model checker based on constrained narrowing currently under construction, publicly available at http://safe-tools.dsic.upv.es/dmc. **DM-Check** uses Maude's **NuITP** [6] inductive theorem prover with a default proof strategy σ as an oracle for two purposes: (i) to check subsumptions $u \mid \varphi \sqsubseteq_{B_\Omega} v \mid \psi$, and (ii) to check whether $u \mid \varphi$ denotes an empty set of states because φ is inductively unsatisfiable. The Maude system modules analyzed by **DM-Check** are topmost rewrite theories $\mathcal{R}$ satisfying the requirements in Theorem 1. At present, only unconditional rewrite theories are supported. This restriction will be removed in a future version.

For the purposes of verifying invariants, **DM-Check** offers the following four commands, where M is the name of the system module specifying $\mathcal{R}$:

1. check-inv in M : $\bigvee_{i \in I} u_i \mid \varphi_i$.
2. check in M : $\bigvee_{i \in I} u_i \mid \varphi_i$ subsumed-by $\bigvee_{j \in J} v_j \mid \psi_j$.
3. intersect in M : $\bigvee_{i \in I} u_i \mid \varphi_i$ with $\bigvee_{j \in J} v_j \mid \psi_j$.
4. add lemma in M : $\bigwedge_{i \in I} u_i = j_i \Rightarrow \bigwedge_{j \in J} u_j = j_j$.

Command (1) uses one step of constrained narrowing to check that $\bigvee_{i \in I} u_i \mid \varphi_i$ is *transition-closed*. It returns any constrained patterns obtained after one-step of constrained narrowing that could not be subsumed into $\bigvee_{i \in I} u_i \mid \varphi_i$ by using a default proof strategy σ as an oracle to prove as many pattern subsumptions as possible and to discard any resulting constrained patterns whose constraints can be shown unsatisfiable.

Command (2) tries to find for each $i \in I$ and $j \in J$ such that $u_i \mid \varphi_i \sqsubseteq_{B_\Omega} v_j \mid \psi_j$ using **NuITP** as an oracle just as the previous command does.

Command (3) symbolically computes the intersection of the sets $\bigcup_{i \in I} [\![u_i \mid \varphi_i]\!]$ and $\bigcup_{j \in J} [\![v_j \mid \psi_j]\!]$ by disjoint B_Ω-unification, again using **NuITP** as an oracle to discard any constrained patterns in the resulting disjunction whose constraints can be shown unsatisfiable.

Command (4) adds an auxiliary lemma to the current module and delegates its proof to **NuITP**.

In all four commands, **DM-Check** and the **NuITP** work in tandem to support the verification methodology proposed in Sect. 3 as follows: (i) **DM-Check** (invoking **NuITP** as an oracle with proof strategy σ^7) supports the *automated* parts of Methodologies (1)–(5). Instead, the **NuITP** supports the *deductive* parts of Methodologies (1)–(5) that could not be proved automatically. We illustrate below this work in tandem between automated and interactive proof by proving several invariants for a fair Readers and Writers protocol example.

---

[7] At present, σ applies a collection of **NuITP** *formula simplification rules*.

DM-Check runs on top of Maude (the latest version, Maude 3.4 or later, should be used). With Maude already installed, to use **DM-Check** one should download it from its web page and unzip the downloaded file into a folder. **DM-Check** already contains **NuITP** as an automatic oracle. However, to carry out interactive proofs with **NuITP** (which also runs on top of Maude) one should download it from its own web page following the instructions there.

Once Maude and **DM-Check** are available, to verify some invariant-related properties one should: (i) start Maude; (ii) load the Maude system module one wants to verify properties about, which should be a topmost rewrite theory, and (iii) give the command `load dm-check-ui.maude` in Maude to load the tool. Once in the tool interface, the commands explained above and illustrated in the example below can be given to verify properties of the given module.

A Fair Readers and Writers Example. To illustrate how the methodology presented in Sect. 3 is supported by **DM-Check** and **NuITP**, we use a fair Readers and Writers protocol whose states have the form $[n] < r, w > [i|j]$, where n is a parameter specifying the maximum number of readers that can participate in the protocol, r is the current number of readers, w is the current number of writers, and the last two components of the state are "token slots" holding i, resp. j, tokens that a reader either: (i) needs to get from the first slot to participate in the protocol; or (ii) returns to the second slot after leaving the protocol. The data elements n, r, w, i and j are all natural numbers with $+$ an associative-commutative natural number constructor with identity element 0. That is, any number is either 0, 1, or a sum $1+ \overset{n}{\ldots} +1$. Readers and writers take turns by "permuting" the two token slots i and j. This makes the system fair, in the sense that neither readers nor writers can be starved from participation. The four transition rules are as follows:

```
rl [w-in]:  [N]< 0,0 >[ 0 | N] => [N]< 0,1 >[0 | N] .
rl [w-out]: [N]< 0,1 >[ 0 | N] => [N]< 0,0 >[N | 0] .
rl [r-in]:  [K + N + M + 1]< N,0 >[M + 1 | K]
         => [K + N + M + 1]< N + 1,0 >[M | K] .
rl [r-out]: [K + N + M + 1]< N + 1,0 >[M | K]
         => [K + N + M + 1]< N,0 >[M | K + 1] .
```

The initial state is *parametric* on the total number of readers. It is defined by the following pattern, where NZ ranges over non-zero natural numbers:

```
[NZ]< 0,0 >[ 0 | NZ ] .
```

Verifying and Inductive Invariant. The proposed inductive invariant is a disjunction of several configurations satisfying two properties: (a) only one writer is allowed in the writing state and (b) $n = r + w + i + j$. In this case, thanks to the expressiveness of the associative-commutative number representation, no explicit constraints are needed in the patterns. This invariant is parametric on the number of readers, since we have variables ranging over numbers in the first argument of the configurations as well as in other arguments, where $N1, N2, M, K$ range over natural numbers and NZ ranges over non-zero natural numbers. The command to verify the inductive invariant, which contains the parametric initial state $[NZ] < 0, 0 > [0|NZ]$ as its first pattern, is as follows:

```
DM-Check> check-inv in R&W-FAIR :
    ([NZ]< 0,0 >[ 0 | NZ]) | true \/
    ([NZ]< 0,1 >[ 0 | NZ]) | true \/
    ([NZ + K + M]< M,0 >[NZ | K]) | true \/
    ([NZ + K + M]< NZ,0 >[M | K]) | true \/
    ([NZ + K + M]< M,0 >[K | NZ]) | true \/
    ([1 + N1 + N2]< 1 + N1, 0 >[0 | N2]) | true \/
    ([1 + N1 + N2]< 0, 0 >[N1 | 1 + N2]) | true \/
    ([1 + N1 + N2]< N1, 0 >[0 | 1 + N2]) | true \/
    ([1 + N1]< 0, 1 >[0 | 1 + N1]) | true \/
    ([1 + N1]< 0, 0 >[1 + N1 | 0]) | true .
```

DM-Check confirms that the inductive invariant is satisfied, so no additional proof obligations need to be handled.

Verifying Deadlock Freedom Positively with the help of NuITP. We can also try to check whether the system is deadlock-free using **DM-Check**. For this, it is enough to prove that the inductive invariant is subsumed by the set of all non-deadlock states, that is, by all states that are *enabled* to make a transition. However, the enabled states have a simple description as a disjunction of patterns, namely, the left-hand sides of the system's rules. The command to check deadlock freedom this way in **DM-Check** is as follows.

```
DM-Check> check in R&W-FAIR :
              ([NZ]< 0,0 >[ 0 | NZ]) | true \/
              ([NZ]< 0,1 >[ 0 | NZ]) | true \/
              ([NZ + K + M]< M,0 >[NZ | K]) | true \/
              ([NZ + K + M]< NZ,0 >[M | K]) | true \/
              ([NZ + K + M]< M,0 >[K | NZ]) | true \/
              ([1 + N1 + N2]< 1 + N1, 0 >[0 | N2]) | true \/
              ([1 + N1 + N2]< 0, 0 >[N1 | 1 + N2]) | true \/
              ([1 + N1 + N2]< N1, 0 >[0 | 1 + N2]) | true \/
              ([1 + N1]< 0, 1 >[0 | 1 + N1]) | true \/
              ([1 + N1]< 0, 0 >[1 + N1 | 0]) | true
subsumed-by
              ((([N]< 0,0 >[ 0 | N]) | true) \/
              (([N]< 0,1 >[ 0 | N]) | true) \/
              (([K + N + M + 1]< N,0 >[M + 1 | K]) | true) \/
              (([K + N + M + 1]< (N + 1), 0 >[M | K]) | true)) .
```

DM-Check indicates that some of the constrained patterns in the inductive invariant cannot be subsumed by any of the patterns specifying the deadlock-free states (i.e., by any transition lefhand side).

```
Term 7: [NZ + K + M]< M, 0 >[NZ | K]
Matching: no matching found
Constraint 7: true

Term 8: [NZ + K + M]< NZ, 0 >[M | K]
Matching: no matching found
Constraint 8: true

Term 9: [NZ + K + M]< M, 0 >[K | NZ]
Matching: no matching found
Constraint 9: true

Term 11: [1 + N1 + N2]< 0, 0 >[N1 | 1 + N2]
```

```
Matching: no matching found
Constraint 11: true

Term 12: [1 + N1 + N2]< N1, 0 >[0 | 1 + N2]
Matching: no matching found
Constraint 12: true
```

The reason is that some of the patterns in the inductive invariant are too general to be subsumed by a single transition lefthand side pattern. But this of course does not mean that they could not be contained in their union, so that the system is indeed deadlock free. This can be shown by passing from the *automatic* verification mode supported by **DM-Check** to the *interactive* mode supported by **NuITP**. As explained in Sect. 3, we can carry out an inductive proof of the containment of the non-subsumed patterns of the inductive invariant into the set of enabled states by first defining the following predicate:

```
op enabled : Conf -> Pred .
eq enabled([N]< 0,0 >[ 0 | N]) = tt .
eq enabled([N]< 0,1 >[ 0 | N]) = tt .
eq enabled([K + N + M + 1]< N,0 >[M + 1 | K]) = tt .
eq enabled([K + N + M + 1]< N + 1,0 >[M | K]) = tt .
```

Then, by using the following **NuITP** commands we can prove that all non-subsumed patterns of the inductive invariant satisfy the **enabled** predicate:

```
NuITP> genset GNAT for Nat is 0 ;; 1 ;; 1 + N:NzNat .
NuITP> genset GNZNAT for NzNat is 1 ;; 1 + N:NzNat .

NuITP> set goal
  (enabled([NZ:NzNat + K:Nat + M:Nat]< M:Nat, 0 >[NZ:NzNat | K:Nat]) = tt) /\
  (enabled([NZ:NzNat + K:Nat + M:Nat]< NZ:NzNat, 0 >[M:Nat | K:Nat]) = tt) .
NuITP> apply cas! to 0 on $3:NzNat .

NuITP> set goal
  enabled([NZ:NzNat + K:Nat + M:Nat]< M:Nat, 0 >[K:Nat | NZ:NzNat]) = tt .
NuITP> apply cas! to 0 on $2:Nat .
NuITP> apply cas! to 0.1 on $1:Nat .

NuITP> set goal enabled([1 + N1:Nat + N2:Nat]< 0, 0 >[N1:Nat | 1 + N2:Nat]) = tt .
NuITP> apply cas! to 0 on $1:Nat .

NuITP> set goal enabled([1 + N1:Nat + N2:Nat]< N1:Nat, 0 >[0 | 1 + N2:Nat]) = tt .
NuITP> apply cas! to 0 on $1:Nat .
```

In this way, the deadlock-freedom invariant is proved for our inductive invariant and therefore for all states reachable from the parametric initial states.

Verifying Mutual Exclusion Negatively. We can prove the mutex invariants negatively. The mutex violation states are characterized by the pattern:

```
[N + 1 + I + 1 + J + L]< N + 1, I + 1 >[L | J]
```

As explained in Sect. 3, we can prove negatively that mutex holds for the inductive invariant, and therefore for the system's reachable states from its parametric set of initial states, by means of the intersection command:

```
DM-Check> intersect in R&W-FAIR :
            ([NZ]< 0,0 >[ 0 | NZ]) | true \/
            ([NZ]< 0,1 >[ 0 | NZ]) | true \/
            ([NZ + K + M]< M,0 >[NZ | K]) | true \/
            ([NZ + K + M]< NZ,0 >[M | K]) | true \/
            ([NZ + K + M]< M,0 >[K | NZ]) | true \/
            ([1 + N1 + N2]< 1 + N1, 0 >[0 | N2]) | true \/
            ([1 + N1 + N2]< 0, 0 >[N1 | 1 + N2]) | true \/
            ([1 + N1 + N2]< N1, 0 >[0 | 1 + N2]) | true \/
            ([1 + N1]< 0, 1 >[0 | 1 + N1]) | true \/
            ([1 + N1]< 0, 0 >[1 + N1 | 0]) | true
 with
            ([N + 1 + I + 1 + J + L]< N + 1,I + 1 >[L | J]) | true .
```

DM-Check confirms an empty intersection, i.e., the mutex invariant holds.

Verifying One-Writer Negatively. We can also prove negatively the *one-writer* invariant, i.e., that there is never more than one writer in the states of the inductive invariant. The pattern specifying the violation of this property is:

```
[N + 1 + I + 1 + J + L]< N,I + 1 + 1 >[L | J]
```

We can prove one-writer negatively by giving the command:

```
DM-Check> intersect in R&W-FAIR :
            ([NZ]< 0,0 >[ 0 | NZ]) | true \/
            ([NZ]< 0,1 >[ 0 | NZ]) | true \/
            ([NZ + K + M]< M,0 >[NZ | K]) | true \/
            ([NZ + K + M]< NZ,0 >[M | K]) | true \/
            ([NZ + K + M]< M,0 >[K | NZ]) | true \/
            ([1 + N1 + N2]< 1 + N1, 0 >[0 | N2]) | true \/
            ([1 + N1 + N2]< 0, 0 >[N1 | 1 + N2]) | true \/
            ([1 + N1 + N2]< N1, 0 >[0 | 1 + N2]) | true \/
            ([1 + N1]< 0, 1 >[0 | 1 + N1]) | true \/
            ([1 + N1]< 0, 0 >[1 + N1 | 0]) | true
 with
            ([N + 1 + I + 1 + J + L]< N,I + 1 + 1 >[L | J]) | true .
```

DM-Check confirms an empty intersection, i.e., the one-writer invariant holds.

Other Examples, besides Readers and Writers fair above and ABP below, can be found in the **DM-Check** web page, including bakery, token ring, an imperative program, and a tree-processing program written in OCaml.

5 A Case Study: The Alternating Bit Protocol

The Alternating Bit Protocol (ABP) [4] is a data layer protocol to achieve reliable communication between two processes over an unreliable channel. For ABP, the *reliable communication property* means that whenever n packets have been delivered, these are the first n packets sent in that particular order. This property has been formally verified using the InvA tool [25]. However, InvA has two limitations that make the verification process much more complex and tedious.

- InvA does not support non-commutative associative operators. However, ABP involves queues that are naturally specified with associative operators.
- InvA only supports invariants specified by Boolean predicates. The user must define auxiliary operators to specify conditions depending on state patterns.

Because of these limitations, the user needs to define and prove many (about 10) lemmas on auxiliary operators to verify ABP with InvA.[8]

This section presents a case study on proving an invariant property of ABP using **DM-Check**. In contrast to the previous work [25], we can freely use any associative operators to specify the system, and enjoy the great expressiveness of constrained patterns to specify invariant conditions. This allows us to easily specify a fairly complex invariant property for reliable communication of ABP, and automatically prove it with only two auxiliary lemmas. The lemmas can also be proved automatically using **NuITP**.

Our ABP specification is adapted from [25], by using associative operators for queues instead of free operators with auxiliary functions. A state of our ABP specification has the form N : B > BPQ || BQ < B' : NL, where N denotes the data currently being sent, B denotes the bit of the sender, BPQ denotes the data channel, BQ denotes the acknowledge channel, B' is the bit of the receiver, and NL denotes the output stream, declared using the following operator:

```
op _:_>_||_<_:_ : iNat Bit BitPacketQue BitQue Bit iList -> Sys [ctor] .
```

The following rewrite rules define the behavior of ABP.[9] They specify the sending of bit-packets over the data channel, the receiving of acknowledgments from the receiver, and the duplication and loss of data in the unreliable channel.

```
vars B B1 B2 : Bit .      var BP  : BitPacket .       var BQ : BitQue .
vars N M : iNat .         var BPQ : BitPacketQue .    var NL : iList .

rl [send-1]: N : B > BPQ || BQ < B1 : NL  =>  N : B > BPQ (B, N) || BQ      < B1 : NL .
rl [send-2]: N : B > BPQ || BQ < B1 : NL  =>  N : B > BPQ           || BQ B1 < B1 : NL .

rl [recv-1a]: N : B   > BPQ || B  BQ < B1 : NL  =>       N : B   > BPQ || BQ < B1 : NL .
rl [recv-1b]: N : on  > BPQ || off BQ < B1 : NL  =>  1 + N : off > BPQ || BQ < B1 : NL .
rl [recv-1c]: N : off > BPQ || on  BQ < B1 : NL  =>  1 + N : on  > BPQ || BQ < B1 : NL .

rl [recv-2a]: N : B > (on,  M) BPQ || BQ < on  : NL  =>  N : B > BPQ || BQ < off : (M :: NL) .
rl [recv-2b]: N : B > (off, M) BPQ || BQ < off : NL  =>  N : B > BPQ || BQ < on  : (M :: NL) .
rl [recv-2c]: N : B > (off, M) BPQ || BQ < on  : NL  =>  N : B > BPQ || BQ < on  : NL .
rl [recv-2d]: N : B > (on,  M) BPQ || BQ < off : NL  =>  N : B > BPQ || BQ < off : NL .

rl [dup-1]: N : B > BP BPQ || BQ    < B1 : NL  =>  N : B > BP BP BPQ || BQ    < B1 : NL .
rl [dup-2]: N : B > BPQ    || B2 BQ < B1 : NL  =>  N : B > BPQ       || B2 B2 BQ < B1 : NL .

rl [drop-1a]: N : B > (off,M) BPQ || BQ < B1 : NL  =>  N : B > BPQ || BQ < B1 : NL .
rl [drop-1b]: N : B > (on,  M) BPQ || BQ < B1 : NL  =>  N : B > BPQ || BQ < B1 : NL .
rl [drop-2a]: N : B > BPQ || off BQ < B1 : NL  =>  N : B > BPQ || BQ < B1 : NL .
rl [drop-2b]: N : B > BPQ || on  BQ < B1 : NL  =>  N : B > BPQ || BQ < B1 : NL .
```

We define the following functions to specify the invariant property for reliable communication. The function aP(*bpQueue, N*) returns True if every packet in

---
[8] For example, in [25] a queue is defined using an empty list nil and a non-associative list constructor _::_, with auxiliary functions such as append defined recursively. Five predicates are declared in [25], and there are many lemmas about interactions between these predicates and the append function. We refer to [25] for further details.

[9] It is possible to simplify some of these rules; for example, drop-1a and drop-1b can be merged into a single rule. However, to ensure a fair comparison, we use the same rewrite rules as [25], except for the use of associative operators.

bpQueue has data value N, defined using three auxiliary functions. The function gen(N) returns the list of the first N numbers in descending order.

```
op aP : BitPacketQue iNat -> iBool .          op n : BitPacket -> iNat .
eq aP(BPQ, N) = allN(ns(BPQ), N) .            eq n((B,N)) = N .

op allN : iSet iNat -> iBool .                eq allN(none, N) = True .
eq allN(N ; NS, N) = allN(NS, N) .            ceq allN(M ; NS, N) = False if (M ~ N) = False .

op gen : iNat -> iList .                      op ns : BitPacketQue -> iSet .
eq gen(0) = 0 .                               eq ns(nil) = none .
eq gen(1 + N) = (1 + N) :: gen(N) .           eq ns(BPQ BP BPQ') = n(BP) ; ns(BPQ BPQ') .
```

We need the following two lemmas for the function `allN` to prove the desired invariant property. This lemma can be automatically proved in **NuITP** using the narrowing induction command.

```
DM-Check> add lemma in ABP-PRED-AUX : (allN(M ; NS, N) = tt) -> (M = N) /\ (allN(NS, N) = tt) .

DM-Check> add lemma in ABP-PRED-AUX : ((tt = allN(M ; ns(BPQ), 1 + N)) /\ ((1 + N :: NL) = (1 + N
                     :: gen(N)))) -> (tt = allN(ns(BPQ), 1 + N)) /\ (M :: NL) = (1 + N :: gen(N)) .
```

Finally, we declare the invariant property for reliable communication of ABP as follows. It is specified as a disjunction of six constrained patterns, each of which indicates that the reliable communication property holds (given by the gen function) and the data and acknowledgment channels are in "good" condition (given by the aP function). **DM-Check** can automatically prove this invariant property, assuming the lemma above.

```
DM-Check> check-inv in ABP-PRED-AUX :
((0 : on > Q' || P < on : nilL) | (aP(Q',0) = tt)) \/
((0 : off > Q' || P < off : nilL) | (aP(Q',0) = tt)) \/
((1 + N : on > Q Q' || P < on : L) | ((1 + N :: L) = gen(1 + N)) /\
                                      (aP(Q,N) = tt) /\ (aP(Q',1 + N) = tt)) \/
((1 + N : off > Q Q' || P < off : L) | ((1 + N :: L) = gen(1 + N)) /\
                                        (aP(Q,N) = tt) /\ (aP(Q',1 + N) = tt)) \/
((N : on > Q || P P' < off : L) | (L = gen(N)) /\ (aP(Q,N) = tt)) \/
((N : off > Q || P P' < on : L) | (L = gen(N)) /\ (aP(Q,N) = tt)) .

 Invariant satisfied using the following lemmas:

 ((tt = allN(M ; ns(BPQ), 1 + N)) /\ (1 + N :: NL) = (1 + N :: gen(N))) ->
 (tt = allN(ns(BPQ), 1 + N)) /\ (M :: NL) = (1 + N :: gen(N))

 tt = allN(M ; NS, N) -> (tt = allN(NS, N)) /\ N = M
```

6 Related Work

The closest to our work is [25] and the Maude Invariant Analyzer tool (**InvA**). They consider some automatic proof search techniques that can be considered as simple theorem proving techniques, now subsumed by **NuITP**: (a) equational simplification, (b) context joinability, (c) unfeasability, and (d) SMT solving (for natural numbers). Indeed, the running example of [25], the alternating bit protocol, is now proved almost automatically. The alternating bit protocol was originally analyzed by [22] and the InvA tool already showed some improvement by automating some deduction steps. In our work, these steps are fully automatic and only two auxiliary lemmas are necessary.

The approach to invariant verification based on proof scores using CafeOBJ [13,21,23] has demonstrated its applicability to many different case studies in the literature. It includes three major steps lemma, case-split, and induction that a human has to encode within the specification and, if all these three major steps are reduced in such a way that they return the expected results, a proof score is achieved. As explained above with the alternating bit protocol, our approach is much more automatic and a user only needs to add an auxiliary lemma. The Invariant Proof Score Generator (IPSG) [28] helps automating some steps such as equational reduction or case splitting but still relies on a human providing the necessary lemmas.

Our work is also related to narrowing-based logical model checking [1–3,8, 10,11], which targets linear temporal logic properties (including invariants). In contrast to our method, they require that the equational theory satisfies the finite variant property or SMT solving is used, which severely limits the systems that can be specified, as mentioned in Sect. 2.

Rubio and Riesco [26] translates Maude rewrite theories into the Microsoft Lean theorem prover. Several properties, including invariants and deadlock, can be proved but the translation does not preserve the modulo reasoning capabilities of Maude and requires heavy user interaction within Lean.

7 Conclusions

We have explained how narrowing-based logical model checking of constrained patterns and inductive theorem proving can be synergistically combined to verify invariants, including inductive ones, of infinite-state systems; and we have used the **DM-Check** and **NuITP** tools and Maude examples to demonstrate this methodology. Much work remains ahead. A richer interaction between **DM-Check** and **NuITP** should be supported in the future. This richer interaction will support, not just invariant verification, but also the reaching of a fixpoint when performing symbolic model checking with **DM-Check**. Furthermore, substantial experimentation with examples is needed to advance both areas.

References

1. Bae, K., Escobar, S., Meseguer, J.: Abstract logical model checking of infinite-state systems using narrowing. In: RTA 2013. LIPIcs, vol. 21, pp. 81–96. Schloss Dagstuhl - Leibniz-Zentrum fuer Informatik (2013)
2. Bae, K., Meseguer, J.: Infinite-state model checking of LTLR formulas using narrowing. In: Proceedings of WRLA 2014. LNCS, vol. 8663, pp. 113–129. Springer, Berlin (2014)
3. Bae, K., Meseguer, J.: Predicate abstraction of rewrite theories. In: RTA-TLCA. Lecture Notes in Computer Science, vol. 8560, pp. 61–76. Springer, Berlin (2014)
4. Bartlett, K.A., Scantlebury, R.A., Wilkinson, P.T.: A note on reliable full-duplex transmission over half-duplex links. Commun. ACM **12**(5), 260–261 (1969)

5. Dershowitz, N., Jouannaud, J.P.: Rewrite systems. In: van Leeuwen, J. (ed.) Handbook of Theoretical Computer Science, vol. B, pp. 243–320. North-Holland (1990)
6. Durán, F., Escobar, S., Meseguer, J., Sapiña, J.: NuITP alpha 21—an inductive theorem prover for maude equational theories. Available at https://nuitp.webs.upv.es/
7. Durán, F., Eker, S., Escobar, S., Martí-Oliet, N., Meseguer, J., Rubio, R., Talcott, C.L.: Programming and symbolic computation in Maude. J. Log. Algebraic Methods Program. **110** (2020)
8. Durán, F., Eker, S., Escobar, S., Martí-Oliet, N., Meseguer, J., Rubio, R., Talcott, C.L.: Equational unification and matching, and symbolic reachability analysis in Maude 3.2. In: Blanchette, J., Kovács, L., Pattinson, D. (eds.) Automated Reasoning—11th International Joint Conference, IJCAR 2022, Haifa, Israel, August 8–10, 2022, Proceedings. Lecture Notes in Computer Science, vol. 13385, pp. 529–540. Springer, Berlin (2022). https://doi.org/10.1007/978-3-031-10769-6_31
9. Escobar, S., Meadows, C., Meseguer, J.: Maude-NPA: cryptographic protocol analysis modulo equational properties. In: Foundations of Security Analysis and Design V, FOSAD 2007/2008/2009 Tutorial Lectures, LNCS, vol. 5705, pp. 1–50. Springer, Berlin (2009)
10. Escobar, S., López-Rueda, R., Sapiña, J.: Symbolic analysis by using folding narrowing with irreducibility and SMT constraints. In: Artho, C., Ölveczky, P.C. (eds.) Proceedings of the 9th ACM SIGPLAN International Workshop on Formal Techniques for Safety-Critical Systems, FTSCS 2023, Cascais, Portugal, 22 October 2023. pp. 14–25. ACM (2023). https://doi.org/10.1145/3623503.3623537
11. Escobar, S., Meseguer, J.: Symbolic model checking of infinite-state systems using narrowing. In: Proceedings of RTA. Lecture Notes in Computer Science, vol. 4533, pp. 153–168 (2007)
12. Escobar, S., Sasse, R., Meseguer, J.: Folding variant narrowing and optimal variant termination. J. Algebraic Logic Program. **81**, 898–928 (2012)
13. Futatsugi, K.: Advances of proof scores in CafeOBJ. Sci. Comput. Program. **224**, 102893 (2022). https://doi.org/10.1016/J.SCICO.2022.102893
14. Goguen, J., Meseguer, J.: Order-sorted algebra I: equational deduction for multiple inheritance, overloading, exceptions and partial operations. Theoret. Comput. Sci. **105**, 217–273 (1992)
15. Lucas, S., Meseguer, J.: Normal forms and normal theories in conditional rewriting. J. Log. Algebr. Meth. Program. **85**(1), 67–97 (2016)
16. Meseguer, J., Thati, P.: Symbolic reachability analysis using narrowing and its application to the verification of cryptographic protocols. J. Higher-Order Symbolic Comput. **20**(1–2), 123–160 (2007)
17. Meseguer, J.: Conditional rewriting logic as a unified model of concurrency. Theoret. Comput. Sci. **96**(1), 73–155 (1992)
18. Meseguer, J.: Membership algebra as a logical framework for equational specification. In: Proceedings of WADT'97. pp. 18–61. Springer LNCS 1376 (1998)
19. Meseguer, J.: Variant-based satisfiability in initial algebras. Sci. Comput. Program. **154**, 3–41 (2018)
20. Meseguer, J.: Generalized rewrite theories, coherence completion, and symbolic methods. J. Log. Algebraic Methods Program. **110** (2020)
21. Ogata, K., Futatsugi, K.: Proof scores in the OTS/CafeOBJ method. In: Najm, E., Nestmann, U., Stevens, P. (eds.) Formal Methods for Open Object-Based Distributed Systems, 6th IFIP WG 6.1 International Conference, FMOODS 2003,

Paris, France, November 19.21, 2003, Proceedings. Lecture Notes in Computer Science, vol. 2884, pp. 170–184. Springer, Berlin (2003). https://doi.org/10.1007/978-3-540-39958-2_12
22. Ogata, K., Futatsugi, K.: Simulation-based verification for invariant properties in the OTS/CafeOBJ method. In: Boiten, E.A., Derrick, J., Smith, G. (eds.) Proceedings of the BCS-FACS Refinement Workshop, REFINE@IFM 2007, Oxford, UK, July 2007. Electronic Notes in Theoretical Computer Science, vol. 201, pp. 127–154. Elsevier (2007). https://doi.org/10.1016/J.ENTCS.2008.02.018
23. Ogata, K., Futatsugi, K.: Theorem proving based on proof scores for rewrite theory specifications of OTSs. In: Iida, S., Meseguer, J., Ogata, K. (eds.) Specification, Algebra, and Software - Essays Dedicated to Kokichi Futatsugi. Lecture Notes in Computer Science, vol. 8373, pp. 630–656. Springer, Berlin (2014). https://doi.org/10.1007/978-3-642-54624-2_31
24. Pnueli, A.: Deduction is forever (1999), invited talk at FM'99 avaliable online at cs.nyu.edu/pnueli/fm99.ps
25. Rocha, C., Meseguer, J.: Mechanical analysis of reliable communication in the alternating bit protocol using the Maude invariant analyzer tool. In: Specification, Algebra, and Software—Essays Dedicated to Kokichi Futatsugi. Lecture Notes in Computer Science, vol. 8373, pp. 603–629. Springer, Berlin (2014)
26. Rubio, R., Riesco, A.: Theorem proving for maude specifications using lean. In: Riesco, A., Zhang, M. (eds.) Formal Methods and Software Engineering—23rd International Conference on Formal Engineering Methods, ICFEM 2022, Madrid, Spain, October 24-27, 2022, Proceedings. Lecture Notes in Computer Science, vol. 13478, pp. 263–280. Springer, Berlin (2022). https://doi.org/10.1007/978-3-031-17244-1_16
27. Slagle, J.R.: Automated theorem-proving for theories with simplifiers commutativity, and associativity. J. ACM **21**(4), 622–642 (1974)
28. Tran, D.D., Ogata, K.: IPSG: invariant proof score generator. In: Leong, H.V., Sarvestani, S.S., Teranishi, Y., Cuzzocrea, A., Kashiwazaki, H., Towey, D., Yang, J., Shahriar, H. (eds.) 46th IEEE Annual Computers, Software, and Applications Conferenc, COMPSAC 2022, Los Alamitos, CA, USA, June 27–July 1, 2022. pp. 1050–1055. IEEE (2022). https://doi.org/10.1109/COMPSAC54236.2022.00164

Time-Bounded Resilience

Tajana Ban Kirigin[1], Jesse Comer[2(✉)], Max Kanovich[3],
Andre Scedrov[2], and Carolyn Talcott[4]

[1] Faculty of Mathematics, University of Rijeka, Rijeka, Croatia
bank@math.uniri.hr
[2] University of Pennsylvania, Philadelphia, PA, USA
jacomer@seas.upenn.edu, scedrov@math.upenn.edu
[3] University College, London, UK
m.kanovich@ucl.ac.uk
[4] SRI International, Menlo Park, CA, USA
carolyn.talcott@sri.com

Abstract. Most research on system design has focused on optimizing *efficiency*. However, insufficient attention has been given to the design of systems optimizing *resilience*, the ability of systems to adapt to unexpected changes or adversarial disruptions. In our prior work, we formalized the intuitive notion of resilience as a property of cyber-physical systems by using a multiset rewriting language with explicit time. In the present paper, we study the computational complexity of a formalization of *time-bounded* resilience problems for the class of η-*simple* progressing planning scenarios, where, intuitively, it is simple to check that a system configuration is critical, and only a bounded number of rules can be applied in a single time step. We show that, in the time-bounded model with n (adversarially-chosen) disruptions, the corresponding time-bounded resilience problem for this class of systems is complete for the Σ_{2n+1}^P class of the polynomial hierarchy, PH. To support the formal models and complexity results, we perform automated experiments for time-bounded verification using the rewriting logic tool Maude.

1 Introduction

Resilience is "the ability of a system to adapt and respond to change (both environmental and internal)" [7]. In recent years, the task of formally defining and analyzing this intuitive notion has drawn interest across domains in computer science, ranging from systems engineering [28,32], particularly cyber-physical systems (CPS) [6,26], to artificial intelligence [14,17,34,36], programming languages [12,20], algorithm design [10,15], and more. Our previous work in [1] was particularly inspired by Vardi's paper [39], in which he articulated a need for computer scientists to reckon with the trade-off between efficiency and resilience.

In [1], we formalized resilience as a property of timed multiset rewriting (MSR) systems [22,24], which are suitable for the specification and verification of various goal-oriented systems. Although the related verification problems are undecidable in general, it was shown that these problems are PSPACE-complete for the class of *balanced* systems, in which facts are of bounded size, and rewrite rules do not change configuration size. A primary challenge in [1] was the formalization of the disruptions against which systems must be resilient. This was achieved by separating the system from the environment, delineating between rules applied by the system and those imposed on the system, such as changes in conditions, regulations, or mission objectives.

Main Contributions. This paper formalizes the notion of *time-bounded* resilience. We focus on the class of η-simple progressing planning scenarios (PPS) and investigate the computational complexity of the corresponding verification problem. Time-bounded resilience is motivated by bounded model checking and automated experiments, which can help system designers verify properties and find counterexamples where their specifications do not satisfy time-bounded resilience. Moreover, bounded versions of resilience problems arise naturally when the missions of the systems being modeled are necessarily bounded at some level. The main contributions of the paper are as follows.

1. We define time-bounded resilience as a property of planning scenarios. Intuitively, a resilient system can accomplish its mission within the given time bounds, even in the presence of a bounded number of disruptions (cf. Definition 11).
2. We investigate the computational complexity of time-bounded resilience problems, showing that for the class of η-simple PPSs with facts of bounded size [23], the time-bounded resilience problem with n updates is complete for the Σ^P_{2n+1} class of the polynomial hierarchy, PH (Corollary 1).
3. We demonstrate that our formalization can be automated, using the rewriting logic tool Maude to perform experiments verifying time-bounded resilience (Sect. 5).

Expository Example. In [1], our study of resilience was motivated by current research into CPSs that perform complex, safety-critical tasks in hostile and unpredictable environments, often autonomously. In this paper, we expand our perspective to consider resilience properties of a broad class of multi-agent systems. For expository purposes, we utilize a running example of a researcher planning travel to attend and present research at a conference. The system rules represent actions of the researcher, while update rules represent travel disruptions and changes to the conference organization. Ultimately, the travel planning process is pointless if the researcher does not arrive at his destination in time for the main event. Consequently, the researcher desires to establish a *resilient* plan, which will allow him to accomplish his goal in spite of some bounded number of disruptions. Details of this planning scenario will be developed throughout Sect. 2, and our Maude implementation in Sect. 5 will be used to analyze its resilience.

Outline. Sect. 2 reviews the timed MSR framework used in Sect. 3 to define time-bounded resilience. In Sect. 4, we investigate the complexity of the verification problem. Section 5 showcases our results on automated verification obtained using Maude. In Sect. 6, we conclude with a discussion of related and future work.

2 Multiset Rewriting Systems

In this section, we review the framework of timed MSR models introduced in our previous work [21,23,24].

2.1 The Rewriting Framework

Terms and Formulas. We fix a finite first-order alphabet Σ with constant, function, and predicate symbols, together with a finite set $\mathcal{B}$ of *base types*. Each constant is associated with a unique base type, and we write Σ_{Cons} to denote the set of all constants in Σ. Each predicate symbol R (resp. function symbol f) is associated with a unique *tuple type* (resp. *arrow type*) $b_1 \times \ldots \times b_k$ (resp. $b_1 \times \ldots \times b_k \to b$), where $b_1, \ldots, b_k, b \in \mathcal{B}$ and k is the arity of R (resp. f). We also assume that Σ contains a special predicate symbol Time with arity zero (more on this later).

We fix sets $\mathcal{V}_{\text{FO}}$ of (first-order) *variables* and $\mathcal{G}$ of *ground constants*, disjoint from each other and from Σ, where each element in $\mathcal{V}_{\text{FO}} \cup \mathcal{G}$ has an associated base type in $\mathcal{B}$. We further assume that $\mathcal{V}_{\text{FO}}$ and $\mathcal{G}$ each contain countably infinitely-many elements associated to each base type. These ground constants will be used to instantiate variables "created" by rewrite rules. *Terms* over Σ are constructed according to the grammar

$$t := x \mid c \mid f(t_1, \ldots, t_k),$$

where x is in $\mathcal{V}_{\text{FO}}$, c is in Σ_{Cons}, f is a function symbol of type $b_1 \times \ldots \times b_k \to b$, and each t_i is a term of type b_i for $i \leq k$ (in which case $f(t_1, \ldots, t_k)$ is a term of type b). *Ground terms* over Σ are constructed similarly:

$$a := d \mid c \mid f(a_1, \ldots, a_k),$$

where d is in $\mathcal{G}$, c is in Σ_{Cons}, f is a function symbol of type $b_1 \times \ldots \times b_k \to b$, and each a_i is a ground term of type b_i for $i \leq k$ (in which case $f(a_1, \ldots, a_k)$ is a ground term of type b). We write $\mathcal{G}_{\text{Terms}}$ for the collection of ground terms over Σ. If R is a predicate symbol of type $b_1 \times \ldots \times b_k$ and $t_1, \ldots, t_k$ are terms of type $b_1, \ldots, b_k$, respectively, then $R(t_1, \ldots, t_k)$ is an *atomic formula*. Similarly, if $a_1, \ldots, a_k$ are ground terms of type $b_1, \ldots, b_k$, respectively, then $R(a_1, \ldots, a_k)$ is an *atomic fact* (or just *fact*).

Modeling Discrete Time. We fix a countably infinite set $\mathcal{V}_{\mathsf{Time}} = \{T_i \mid i \in \mathbb{N}\}$ of *time variables*. *Timestamped atomic formulas* are of the form $F@(T+D)$, where F is an atomic formula, T is a time variable, and D is a natural number; note that if $D=0$, we prefer to write $F@T$ instead of $F@(T+0)$. *Timestamped facts* are of the form $F@t$, where F is an atomic fact and $t \in \mathbb{N}$ is its *timestamp*. For brevity, we frequently refer to timestamped facts simply as facts. Clearly, given a timestamped atomic formula $F@(T+D)$, we can obtain a timestamped fact $G@t$ by uniformly substituting ground terms for variables in F and setting $t = N + D$ for some natural number N.

Configurations and Rewrite Rules. *Configurations* are multisets of timestamped facts $\mathcal{S} = \{\mathsf{Time}@t, F_1@t_1, \ldots, F_n@t_n\}$ with exactly one occurrence of a Time fact whose timestamp t is the *global time* in $\mathcal{S}$. We write Values($\mathcal{S}$) to denote the set of all ground terms and timestamps occurring in $\mathcal{S}$. Configurations are modified by *multiset rewrite rules*. Only one rule, Tick, modifies global time:

$$\mathsf{Time}@T \longrightarrow \mathsf{Time}@(T+1) \tag{1}$$

where T is a time variable. The Tick rule modifies a configuration to which it is applied by advancing the global time by one. The remaining rules are *instantaneous* in that they do not modify the global time but may modify the remaining facts of a configuration. Instantaneous rules are given by expressions of the form

$$\begin{aligned}\mathsf{Time}@T, \mathcal{W}, F_1@T_1, \ldots F_n@T_n \mid \mathcal{C} \\ \longrightarrow \mathsf{Time}@T, \mathcal{W}, Q_1@(T+D_1), \ldots Q_m@(T+D_m)\end{aligned} \tag{2}$$

where $\mathcal{W}$ (the *side condition* of the rule) is a multiset of timestamped atomic formulas, $F_i@T_i$ is a timestamped atomic formula for each $i \leq n$, and $Q_j@(T+D_j)$ is a timestamped atomic formula for each $j \leq m$. The *precondition* of the rule is the multiset $\{\mathsf{Time}@T\} \cup \mathcal{W} \cup \{F_i@T_i \mid i \leq n\}$, while its *postcondition* is the multiset $\{\mathsf{Time}@T\} \cup \mathcal{W} \cup \{Q_j@(T+D_j) \mid j \leq m\}$. We require that no atomic formula in the multiset $\{F_i@T_i \mid i \leq n\}$ appears with the same multiplicity as it appears in the multiset $\{Q_j@(T+D_j) \mid j \leq m\}$. Furthermore, no timestamped atomic formulas containing the predicate Time can occur in $\{F_i@T_i \mid i \leq n\} \cup \{Q_j@(T+D_j) \mid j \leq m\}$. The *guard* $\mathcal{C}$ of the rule is a set of *time constraints* of the form

$$T_1 > T_2 \pm N \quad \text{or} \quad T_1 = T_2 \pm N,$$

where T_1 and T_2 are time variables and $N \subset \mathbb{N}$ is a natural number; all constraints in $\mathcal{C}$ must involve only the time variables occurring in the rule's precondition.

A *ground substitution* is a partial map $\sigma : \mathcal{V}_{\mathsf{FO}} \cup \mathcal{V}_{\mathsf{Time}} \to \mathcal{G}_{\mathsf{Terms}} \cup \mathbb{N}$ which maps first-order variables to ground terms and time variables to natural numbers. Given a multiset W of timestamped atomic formulas, we write $W\sigma$ to denote the multiset of timestamped facts obtained by simultaneously substituting all first-order variables and time variables in W with their image under σ. Given a set $\mathcal{C}$ of time constraints with time variables from W, we say that $\mathcal{C}\sigma$ is *satisfied* if each time constraint in $\mathcal{C}$ evaluates to true for the substituted timestamps. Given a multiset W of timestamped atomic formulas, we write $\mathrm{Var}(W)$ to denote the set of first-order variables and time variables occurring in W. Given an instantaneous rule r given by $W \mid \mathcal{C} \longrightarrow W'$, we write $\mathrm{Fresh}(r)$ to denote the set $\mathrm{Var}(W') \setminus \mathrm{Var}(W)$.

A ground substitution matching an instantaneous rule r given by $W \mid \mathcal{C} \longrightarrow W'$ to a configuration $\mathcal{S}$ is a ground substitution σ with $\mathrm{dom}(\sigma) = \mathrm{Var}(W \cup W')$ such that every element of $\mathrm{Var}(W)$ is mapped to an element in $\mathrm{Values}(\mathcal{S})$, and the restriction of σ to $\mathrm{Fresh}(r)$ is an injective map whose range is contained in $\mathcal{G} \setminus \mathrm{Values}(\mathcal{S})$. In other words, σ assigns first-order variables (resp. time variables) occurring in W to ground terms (resp. timestamps) occurring in $\mathcal{S}$, and each distinct first-order variable in $\mathrm{Fresh}(r)$ to a *fresh* ground constant which does not occur in $\mathcal{S}$.

An instantaneous rule r given by $W \mid \mathcal{C} \longrightarrow W'$ is *applicable* to a configuration $\mathcal{S}$ if there exists a ground substitution matching r to $\mathcal{S}$ such that $W\sigma \subseteq \mathcal{S}$ and $\mathcal{C}\sigma$ is satisfied; in this case, we refer to the expression $r\sigma$ given by $W\sigma \mid \mathcal{C}\sigma \longrightarrow W'\sigma$ as an *instance* of the rule r. The result of applying the rule instance $r\sigma$ to $\mathcal{S}$ is the configuration $(\mathcal{S} \setminus W\sigma) \cup W'\sigma$. If $\mathcal{W}$ is the side condition of r, and T is the global time in $\mathcal{S}$, then we say that the timestamped facts occurring in $(W \setminus (\mathcal{W} \cup \{\mathsf{Time}@T\}))\sigma$ are *consumed*, while those in $(W' \setminus (\mathcal{W} \cup \{\mathsf{Time}@T\}))\sigma$ are *created*. Note that a fact for the predicate Time is never created by an instantaneous rule. We write $\mathcal{S} \longrightarrow_r \mathcal{S}'$ for the one-step relation where the configuration $\mathcal{S}$ is rewritten to $\mathcal{S}'$ using an instance of the rule r. It is worth emphasizing, at this point, that configurations are *grounded*, while rewrite rules are *symbolic*.

Some Examples. We now give some examples to elucidate our formalism. Consider the alphabet containing the predicate symbols Time, At, Event, Attended, and Flight_D (where $D \in \{1, \ldots, 12\}$), and the constant symbols no, done, main, airport, center, id_{14}, and id_{215}. Recall that, in our expository example, we are modeling a researcher with a goal of traveling to attend a conference. We interpret the timestamped atomic formula $\mathsf{Flight}_D(id, c_1, c_2)@T$ to mean that the flight with flight id id from city c_1 to city c_2 departs at time T and has a duration of approximately D hours.

The timestamped fact $\mathsf{At}(\mathsf{FRA}, \mathsf{center})@0$ is interpreted to mean that the researcher is at Frankfurt city center at the initial time step 0. For this scenario, each time step is interpreted as the passage of one minute. For ease of readability, we adopt a more convenient representation of timestamps, with 0 denoting midnight on the initial day of the planning scenario. Then, we write $\mathsf{Time}@(3d\ 14{:}42)$ to indicate that the current time is 14:42 on the 3rd day of the scenario. We do this is in lieu of writing the more burdensome timestamp $\mathsf{Time}@5202$. The fact $\mathsf{Event}(\mathsf{main}, \mathsf{id}_{215})@(5d\ 12{:}00)$ specifies that the main event of the conference, with event identifier 215, will take place at noon on the 5th

day. Bringing this all together, consider the following configuration.

$$\{\text{Time}@(3d\ 14{:}42), \underline{\text{Attended}}(\text{main}, \text{no})@0,\ \text{At}(\text{FRA}, \text{airport})@(3d\ 14{:}05), \\ \underline{\text{Event}}(\text{main}, \text{id}_{215})@(5d\ 12{:}00), \text{Flight}_2(\text{id}_{14}, \text{FRA}, \text{DBV})@(3d\ 15{:}25)\} \quad (3)$$

This configuration describes a state of the system. The time is 14:42 on the 3rd day of the scenario, and the researcher arrived at Frankfurt airport (FRA) at 14:05. The main event of the conference is at noon in two days in Dubrovnik (DBV), and has, obviously, not yet been attended by the researcher. Flight id_{14} is a direct flight from Frankfurt to Dubrovnik, which departs at 15:25 and has a duration of approximately two hours.

In addition to modeling states of the system via configurations, we also want our formalism to be able to model actions taken by the researcher, such as boarding a given flight. To this end, consider the rule

$$\text{Time}@T, \text{Flight}_2(a,x,y)@T_1, \text{At}(x,\text{airport})@T_2,\ |\ T = T_1, T_2 + 30 \leq T \\ \longrightarrow \text{Time}@T, \text{Flight}_2(a,x,y)@T_1, \text{At}(y,\text{airport})@(T+120), \quad (4)$$

with side condition $\{\text{Flight}_2(a,x,y)@T_1\}$. This rule means that if the departure time of a two-hour flight with flight id a from city x to city y will depart at time T, and the researcher is at the airport in city x at time T_2, where T_2 is at least 30 minutes prior to T, then he can take the flight, arriving at the airport in city y after two hours.

Note that the rule (Eq. 4) is not applicable to the configuration (Eq. 3). In particular, the time constraint $T = T_1$ cannot be satisfied by any ground assignment for the rule to the configuration. However, rule (Eq. 4) *is* applicable to the configuration resulting from the successive application of 43 Tick rules to configuration (Eq. 3), which results in the same configuration, except with the timestamp for Time updated to $(3d\ 15{:}25)$ (i.e., the departure time of the flight). Then the ground substitution σ given by

$$\sigma(T) = 3d\ 15{:}25 \qquad\qquad \sigma(a) = \text{id}_{14}$$
$$\sigma(T_1) = 3d\ 15{:}25 \qquad\qquad \sigma(x) = \text{FRA}$$
$$\sigma(T_2) = 3d\ 14{:}05 \qquad\qquad \sigma(y) = \text{DBV}$$

applied to the rule (Eq. 4) yields an instance which can be applied to configuration (Eq. 3), resulting in the following configuration:

$$\{\text{Time}@(3d\ 15{:}25), \underline{\text{Attended}}(\text{main}, \text{no})@0,\ \text{At}(\text{DBV}, \text{airport})@(3d\ 17{:}25), \\ \underline{\text{Event}}(\text{main}, \text{id}_{215})@(5d\ 12{:}00), \text{Flight}_2(\text{id}_{14}, \text{FRA}, \text{DBV})@(3d\ 15{:}25)\}. \quad (5)$$

Timed MSR Systems. We now turn to the timed MSR systems introduced in [24].

Definition 1. *A timed MSR system* $\mathcal{A}$ *is a set of rules containing only instantaneous rules (Eq. 2) and the Tick rule (Eq. 1).*

A sequence of consecutive rule applications represents an execution or process within the system. A *trace* of timed MSR rules $\mathcal{A}$ starting from an initial configuration $\mathcal{S}_0$ is a sequence of configurations: $\mathcal{S}_0 \longrightarrow \mathcal{S}_1 \longrightarrow \mathcal{S}_2 \longrightarrow \cdots \longrightarrow \mathcal{S}_n$, such that for all $0 \leq i \leq n-1$, $\mathcal{S}_i \longrightarrow_{r_i} \mathcal{S}_{i+1}$ for some $r_i \in \mathcal{A}$. For our complexity results, we assume traces are annotated with the rule instances used to obtain the next configuration in the trace, so valid traces can be recognized in polynomial time (cf. Remark 4).

Reachability problems for MSR systems are to determine whether or not a trace from some initial configuration to some specified configuration exists. In general, these problems are often undecidable [24], and so restrictions are imposed in order to obtain decidability[1]. In particular, we use MSR systems with only *balanced* rules.

Definition 2. (*Balanced Rules*, [24]) A timed MSR rule is *balanced* if the numbers of facts on left and right sides of the rule are equal.

Systems containing only balanced rules represent an important class of *balanced systems*, for which several reachability problems have been shown to be decidable [23]. Balanced systems are suitable, *e.g.*, for modeling scenarios with a fixed amount of total memory. Balanced systems have the following important property:

Proposition 1. ([23]) *Let $\mathcal{R}$ be a set of balanced rules. Let $\mathcal{S}_0$ be a configuration with exactly m facts (counting multiplicities). Let $\mathcal{S}_0 \longrightarrow \cdots \longrightarrow \mathcal{S}_n$ be an arbitrary trace of $\mathcal{R}$ rules starting from $\mathcal{S}_0$. Then for all $0 \leq i \leq n$, $\mathcal{S}_i$ has exactly m facts.*

2.2 Progressing Timed Systems

In this section, we review a particular class of timed MSR systems, called *progressing timed MSR systems* (PTSs) [21,22], in which only a bounded number of rules can be applied in a single time step. This is a natural condition, similar to the *finite-variability assumption* used in the temporal logic and timed automata literature [18].

Definition 3. (*Progressing Timed System*, [21]) An instantaneous rule r of the form in (Eq. 2) is *progressing* if the following all hold: (i) $n = m$ (i.e., r is balanced); (ii) r consumes *only* facts with timestamps in the past or at the current time, *i.e.*, in (Eq. 2), the set of constraints $\mathcal{C}$ of r contains the set $\mathcal{C}_r = \{\, T \geq T_i \mid F_i@T_i, 1 \leq i \leq n\,\}$; (iii) r creates *at least one* fact with timestamp greater than the global time, *i.e.*, in (Eq. 2), $D_i \geq 1$ for at least one $i \in \{1, \ldots, n\}$. A timed MSR system $\mathcal{A}$ is a *progressing timed MSR system (PTS)* if all instantaneous rules of $\mathcal{A}$ are progressing.

---

[1] For a discussion of various conditions in the model that may affect complexity, see [23,24].

Note that the rule (Eq. 4) is progressing. A timestamped fact in a configuration $\mathcal{S}$ is a *future fact* if its timestamp is strictly greater than the timestamp of the Time@t fact in $\mathcal{S}$. Future facts are "not available" in the sense that they cannot be consumed by a progressing rule before a sufficient number of Tick rule applications.

Remark 1. For readability, we assume the set of constraints for all rules r, contains the set $\mathcal{C}_r$, as in Definition 3, and do not always write $\mathcal{C}_r$ explicitly.

2.3 Timed MSR for the Specification of Resilient Systems

We now review additional notation for the purpose of specifying resilience, as introduced in [1]. The resilience framework divides the system from an external entity, such as the environment, regulatory authorities, or an adversary. We model various types of disruptive changes to the system state or goals.

Definition 4. (*Planning Configuration*, [1]) Let $\Sigma_P = \Sigma_G \uplus \Sigma_C \uplus \Sigma_S \uplus \{\text{Time}\}$ consist of four pairwise disjoint sets of predicate symbols, Σ_G, Σ_C, Σ_S and $\{\text{Time}\}$. Facts constructed using predicates from Σ_G are called *goal facts*, from Σ_C *critical facts*, and from Σ_S *system facts*. Facts constructed using predicates from $\Sigma_C \cup \Sigma_G$ are called *planning facts*. Configurations over Σ_P predicates are called *planning configurations*.

For readability, we underline predicates in planning facts and refer to planning configurations as configurations for short. The behavior of the system is represented by traces of MSR rules. A system should achieve its goal while not violating predetermined critical conditions. This is made precise in the following two definitions.

Definition 5. (*Critical/Goal Configurations*, [1]) A *critical* (resp. *goal*) *configuration specification* $\mathcal{CS}$ (resp. $\mathcal{GS}$) is a set of pairs $\{\langle \mathcal{S}_1, \mathcal{C}_1 \rangle, \ldots, \langle \mathcal{S}_n, \mathcal{C}_n \rangle\}$, with each pair $\langle \mathcal{S}_j, \mathcal{C}_j \rangle$ being of the form $\langle \{F_1 @ T_1, \ldots, F_{p_j} @ T_{p_j}\}, \mathcal{C}_j \rangle$, where $T_1, \ldots, T_{p_j}$ are time variables, $W = \{F_1, \ldots, F_{p_j}\}$ is a multiset of timestamped atomic formulas, with at least one occurrence of a critical (resp. goal) predicate symbol, and $\mathcal{C}_j$ is a set of time constraints involving only variables $T_1, \ldots, T_{p_j}$. A configuration $\mathcal{S}$ is a *critical configuration* w.r.t. $\mathcal{CS}$ (resp. a *goal configuration* w.r.t. $\mathcal{GS}$) if for some $1 \leq i \leq n$, there is a grounding substitution σ with $\text{dom}(\sigma) = \text{Var}(W)$ such that $\mathcal{S}_i \sigma \subseteq \mathcal{S}$ and $\mathcal{C}_i \sigma$ is satisfied.

Definition 6. (*Compliant Traces*, [1]) A trace is *compliant* with respect to a critical configuration specification $\mathcal{CS}$ if it does not contain any critical configuration w.r.t. $\mathcal{CS}$.

Note that critical configuration specifications and goal configuration specifications, like rewrite rules, are *symbolic*. Reaching a critical configuration may be interpreted as a *safety violation*, while a compliant trace may be interpreted as a *safe trace*. As an example, suppose that in the example alphabet introduced earlier, the predicate symbol <u>Attended</u> is in Σ_C, while the predicate symbol <u>Event</u> is in Σ_G. Then the goal configuration specification

$$\{\langle\{\underline{\mathsf{Attended}}(\mathsf{main},\mathsf{done})@T_1,\underline{\mathsf{Event}}(\mathsf{main},x)@T_2\},\emptyset\rangle\}$$

indicates that the main event should be attended, while the critical configuration specification

$$\{\langle\,\mathsf{Time}@T,\underline{\mathsf{Attended}}(\mathsf{main},\mathsf{no})@T_1,\underline{\mathsf{Event}}(\mathsf{main},x)@T_2\},\{T>T_2\}\rangle\}$$

denotes that it is critical not to participate in the main event.

Definition 7. (*System Rules and Update Rules,* [1]) Fix a planning alphabet Σ_P. A *system rule* is either the Tick rule (Eq. 1) or a rule of form in (Eq. 2) which does not consume or create planning facts. An *update rule* is an instantaneous rule that is of one of the following types: (a) a *system update rule* (SUR) such that planning facts can only occur in the side condition of the rule; or (b) a *goal update rule* (GUR) that either consumes or creates at least one goal fact and such that critical facts can only occur in the side condition of the rule.

For example, the following system rule specifies that the traveler needs 40 minutes to get from the departing city center to the airport:

$$\mathsf{Time}@T,\ \mathsf{At}(x,\mathsf{center})@T_1\ |\ T_1 \leq T$$
$$\longrightarrow\ \mathsf{Time}@T,\mathsf{At}(x,\mathsf{airport})@(T+40).$$

The rule (Eq. 4) is another example of a system rule. System rules specify the behavior of the system, while disruptions are modeled via update rules. Intuitively, GUR model external interventions in the system, such as mission changes, additional tasks, etc., while SUR model changes in the system that are not due to the intentions of the system's agents, *e.g.*, technical errors or malfunctions such as flight delays. Both goal and system update rules can create and/or consume system facts, which technically simplifies modeling the impact of changes on the system and its response. For example, the following GUR models a change in the scheduled time of the main event.

$$\mathsf{Time}@T,\ \underline{\mathsf{Event}}(\mathsf{main},x)@T_1,\ \longrightarrow\ \mathsf{Time}@T,\underline{\mathsf{Event}}(\mathsf{main},x)@(T+60),$$

while the following SUR models a 30 min flight delay:

$$\mathsf{Time}@T,\mathsf{Flight}_D(a,x,y)@T_1\ \longrightarrow\ \mathsf{Time}@T,\mathsf{Flight}_D(a,x,y)@(T+30).$$

Definition 8. (*Planning Scenario,* [1]) If $\mathcal{R}$ and $\mathcal{E}$ are sets of system and update rules, $\mathcal{GS}$ and $\mathcal{CS}$ are a goal and critical configuration specifications, and $\mathcal{S}_0$ is an initial configuration, then the tuple $(\mathcal{R},\mathcal{GS},\mathcal{CS},\mathcal{E},\mathcal{S}_0)$ is a *planning scenario*.

Definition 9. (*Progressing Planning Scenario (PPS)*) We say that a planning scenario $(\mathcal{R},\mathcal{GS},\mathcal{CS},\mathcal{E},\mathcal{S}_0)$ is *progressing* if all rules in $\mathcal{R}$ and $\mathcal{E}$ are progressing.

The progressing condition in Definition 9 implies a bound on the number of rules that can be applied in a single unit of time (cf. Proposition 2). We also assume an upper-bound on the size of facts allowed to occur in traces, where the size of a timestamped fact $F@t$ is the number of symbols from Σ occurring in F, counting repetitions. Without this bound (among other restrictions), many interesting decision problems are undecidable [13,23]. We also confine attention to classes of η-simple PPSs.

Definition 10. Let η denote a fixed positive integer. We say that a planning scenario $A = (\mathcal{R}, \mathcal{GS}, \mathcal{CS}, \mathcal{E}, \mathcal{S}_0)$ is η-simple if the total number of variables (including both first-order and time variables) appearing in each pair $\langle \mathcal{S}_i, \mathcal{C}_i \rangle$ in $\mathcal{CS}$ is less than η.

For every planning scenario $A = (\mathcal{R}, \mathcal{GS}, \mathcal{CS}, \mathcal{E}, \mathcal{S}_0)$, there exists some least η such that A is η-simple; intuitively, this η is a measure of the complexity of verifying compliance of traces with respect to $\mathcal{CS}$. Proposition 4 in Sect. 4 makes this intuition precise.

Remark 2. By inspecting the rules and the critical configuration specification, it is easy to check that our expository travel example is 3-simple and progressing.

3 Time-Bounded Resilience Verification Problems

In this section, we formalize time-bounded resilience as a property of planning scenarios. Intuitively, we want to capture the notion of a system which can achieve its goal within a fixed amount of time, despite the application of up to n instances of update rules. An initial idea might be to require that the system can achieve its goal in the allotted time, regardless of when updates are applied. However, this is too restrictive: many systems will fail to achieve their goal in the face of adversarial actions which can be applied arbitrarily often. Instead, the system will have $a + b$ time units to achieve its goal, and update rules can only be applied in the first a time steps; the last b time steps are the *recovery time* afforded to the system.

Definition 11. (*The (n, a, b)-resilience problem*) Let $a \in \mathbb{Z}^+$ and $b \in \mathbb{N}$. We define (n, a, b)-resilience by recursion on n. Inputs to the problem are planning scenarios $A = (\mathcal{R}, \mathcal{GS}, \mathcal{CS}, \mathcal{E}, \mathcal{S}_0)$. A trace is $(0, a, b)$-*resilient with respect to A* if it is a compliant trace of $\mathcal{R}$ rules from $\mathcal{S}_0$ to a goal configuration and contains at most $a + b$ applications of the Tick rule. Let t_0 denote the global time in the configuration $\mathcal{S}_0$. For $n > 0$, a trace τ is (n, a, b)-*resilient with respect to A* if

1. τ is $(0, a, b)$-resilient with respect to A, and
2. for any system or goal update rule $r \in \mathcal{E}$ such that $\mathcal{S}_i \longrightarrow_r \mathcal{S}'_{i+1}$ for some configuration $\mathcal{S}_i$ in τ with global time t_i, where $d_i = t_i - t_0 \leq a$, there is a *reaction trace* τ' of $\mathcal{R}$ rules from $\mathcal{S}'_{i+1}$ to a goal configuration $\mathcal{S}'$ such that τ' is $(n - 1, a - d_i, b)$-resilient with respect to $A' = (\mathcal{R}, \mathcal{GS}, \mathcal{CS}, \mathcal{E}, \mathcal{S}'_{i+1})$.

A planning scenario $A = (\mathcal{R}, \mathcal{GS}, \mathcal{CS}, \mathcal{E}, \mathcal{S}_0)$ is (n, a, b)-*resilient* if an (n, a, b)-resilient trace with respect to A exists. The (n, a, b)-*resilience problem* is to determine if a given planning scenario A is (n, a, b)-resilient.

Figure 1 provides a visual depiction of Definition 11.

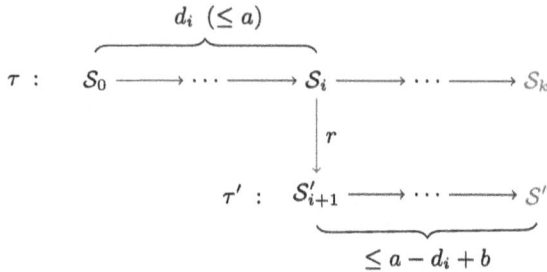

Fig. 1. An (n, a, b)-resilient trace τ and an $(n-1, a-d_i, b)$-resilient reaction trace τ'. The horizontal arrows correspond to system rule applications, while the downward-facing arrow represents an update rule application. The configurations $\mathcal{S}_k$ and $\mathcal{S}'$ on the far right are goal configurations

The reaction trace τ' in Definition 11 can be interpreted as a change in the plan τ, made in response to an external disruption (i.e., the system/goal update rule r) imposed on the system. Note that it is this "replanning" aspect of our definition that intuitively distinguishes it from the related notion of robustness.

Remark 3. In Definition 11, the global time t' in $\mathcal{S}'$ satisfies $t' - t_0 \leq a + b$; i.e., despite the application of n instances of update rules, an (n, a, b)-resilient trace reaches a goal within $a + b$ time units. Furthermore, observe that a trace is (n, a, b)-resilient with respect to a planning scenario A if and only if it is (n, a, b')-resilient with respect to A for all $b' \geq b$. Similarly, all (n, a, b)-resilient traces with respect to A are (n', a, b)-resilient with respect to A for all $n' \leq n$.

It is worthwhile to note that Definition 11 can be seen as a modification of [1, Definitions 9–10], in which (*i*) we include the parameters a and b, (*ii*) we consider both system and goal update rules simultaneously, and (*iii*) the *recoverability condition*, which is not mentioned in this work, is the total relation on configurations of A.

4 Computational Complexity of Time-Bounded Resilience

In this section, we state and prove our results on the computational complexity of the time-bounded resilience problem defined in Sect. 3. To see this, we first state a known bound on the number of instances of instantaneous rules appearing between two consecutive instances of *Tick* rules in a trace of only progressing rules.

Proposition 2. ([21]) *Let $\mathcal{R}$ be a set of progressing rules, $\mathcal{S}_0$ an initial configuration, and m the number of facts in $\mathcal{S}_0$. For all traces τ of $\mathcal{R}$ rules starting from $\mathcal{S}_0$, let*

$$\mathcal{S}_i \longrightarrow_{Tick} \mathcal{S}_{i+1} \longrightarrow \cdots \longrightarrow \mathcal{S}_j \longrightarrow_{Tick} \mathcal{S}_{j+1}$$

be any subtrace of τ with exactly two instances of the Tick rule, one at the beginning and the other at the end. Then $j - i \leq m$.

Proposition 2 guarantees that the lengths of (n, a, b)-resilient traces of a PPS A are polynomially-bounded in the size of the input representation of A.

Proposition 3. *Let $A = (\mathcal{R}, \mathcal{GS}, \mathcal{CS}, \mathcal{E}, \mathcal{S}_0)$ be a PPS and m be the number of facts in $\mathcal{S}_0$. Then the length of any (n, a, b)-resilient trace of A is bounded by $(a + b + 1)m$.*

In preparation for our (n, a, b)-resilience upper bound result (Theorem 1), we now turn to the complexity of some fundamental decision problems pertaining to planning scenarios. We only state the problems and their complexity for η-simple PPSs with facts of bounded size; more detail can be found in the technical report [3].

Definition 12. *The goal (resp. critical) recognition problem is to determine, given a planning scenario $A = (\mathcal{R}, \mathcal{GS}, \mathcal{CS}, \mathcal{E}, \mathcal{S}_0)$ and a configuration $\mathcal{S}$, whether or not $\mathcal{S}$ is a goal (resp. critical) configuration w.r.t. $\mathcal{GS}$ (resp. $\mathcal{CS}$); cf. Definition 5.*

More broadly, we are interested in checking trace compliance.

Definition 13. *The trace compliance problem is to determine, given a planning scenario $A = (\mathcal{R}, \mathcal{GS}, \mathcal{CS}, \mathcal{E}, \mathcal{S}_0)$ and a trace τ of $\mathcal{R}$-rules starting from $\mathcal{S}_0$, whether or not τ is compliant w.r.t. $\mathcal{CS}$ (cf. Definition 6).*

They key observation, and the one underlying our restriction to η-simple PPSs (cf. Definition 10), is that the trace compliance problem is tractable for this class.

Proposition 4. *For η-simple planning scenarios, the trace compliance problem is in P.*

Remark 4. If $A = (\mathcal{R}, \mathcal{GS}, \mathcal{CS}, \mathcal{E}, \mathcal{S}_0)$ is an η-simple PPS, and $\mathcal{S}$ a configuration with the same number of facts as $\mathcal{S}_0$, then given an appropriate ground substitution, we can verify in polynomial time in the size of A that $\mathcal{S}$ is a goal configuration w.r.t. $\mathcal{GS}$. In the proof of Theorem 1, for ease of exposition, we will also assume that these ground substitutions come with a pointer to the appropriate pair $\langle \mathcal{S}_i, \mathcal{C}_i \rangle$ in $\mathcal{GS}$ to which the substitution should be applied. Furthermore, given an arbitrary PPS $A = (\mathcal{R}, \mathcal{GS}, \mathcal{CS}, \mathcal{E}, \mathcal{S}_0)$, if $\mathcal{S}$ is a configuration with the same number of facts as $\mathcal{S}_0$, then given an appropriate ground substitution, we can verify in polynomial time in the size of A that $\mathcal{S}$ is a goal configuration w.r.t. $\mathcal{CS}$. Similarly, given an appropriate ground substitution, we can verify in polynomial time in the size of A that a rule r is applicable to $\mathcal{S}$, and whether or not $\mathcal{S}'$ is the result of this application.

We now turn our attention to the computational complexity of the (n, a, b)-resilience problem. To establish our complexity results, we will utilize the quantifier-alternation characterization of PH (cf. [2,35,38]), according to which a decision problem is in Σ_n^P (for n odd) if and only if there exists a polynomial-time algorithm M such that an input x is a *yes* instance of the problem if and only if

$$\exists u_1 \forall u_2 \exists u_3 \ldots \forall u_{n-1} \exists u_n\ M(x, u_1, \ldots, u_n) \text{ accepts,}$$

where the u_i are polynomially-bounded in the size of x. We now establish an upper bound on the complexity of the (n, a, b)-resilience problem (cf. Definition 11).

Theorem 1. *For η-simple PPSs with traces containing only facts of bounded size and all $a \in \mathbb{Z}^+$ and $n, b \in \mathbb{N}$, there exists a decision procedure of complexity Σ_{2n+1}^P for the (n, a, b)-resilience problem.*

Proof. We show by induction on n that, for each $a \in \mathbb{Z}^+$ and $b \in \mathbb{N}$, there exists a polynomial-time algorithm $M_n^{a,b}$ such that an η-simple PPS $\mathcal{A} = (\mathcal{R}, \mathcal{GS}, \mathcal{CS}, \mathcal{E}, \mathcal{S}_0)$ is (n, a, b)-resilient if and only if

$$\exists \mathcal{T}_0 \forall \rho_1 \exists \mathcal{T}_1 \ldots \forall \rho_n \exists \mathcal{T}_n\ M_n^{a,b}(\mathcal{A}, \mathcal{T}_0, \mathcal{T}_1, \ldots, \mathcal{T}_n, \rho_1, \ldots, \rho_n) \text{ accepts.} \quad (6)$$

The existentially quantified variables $\mathcal{T}_i$ range over triples of the form (τ_i, σ_i, j_i), where τ_i is a trace of $\mathcal{R}$-rules and σ_i is a ground substitution from $\langle \mathcal{S}_{j_i}, C_{j_i} \rangle$ in $\mathcal{GS}$ to the last configuration of τ_i. By Proposition 3, such a witness $\mathcal{T}_i$ is polynomially-bounded in the size of the input representation of $\mathcal{A}$. The universally quantified variables ρ_i range over triples of the form (r_i, σ_i, j_i), where $r_i \in \mathcal{E}$ and σ_i is a ground substitution from the j_i^{th} configuration of τ_{i-1} to the first configuration of τ_i. The witnesses ρ_i are also clearly polynomially-bounded in the size of the input representation of $\mathcal{A}$.

For the base case, the algorithm $M_0^{a,b}$ first verifies that $\mathcal{A}$ meets the syntactic requirements of an η-simple PPS. If so, then we verify, given $\mathcal{T}_0 = (\tau_0, \sigma_0, j_0)$, that τ_0 has at most $a + b$ applications of the *Tick* rule, is compliant, and leads to a goal. By Proposition 4, since τ_0 is polynomially-bounded in $\mathcal{A}$, we can verify compliance of τ_0 in polynomial time in $\mathcal{A}$. By Remark 4, we can verify in polynomial time in $\mathcal{A}$, given (σ_0, j_0), that the last configuration of τ_0 is a goal. Hence $M_0^{a,b}(\mathcal{A}, \mathcal{T}_0)$ runs in polynomial time, and $\mathcal{A}$ is $(0, a, b)$-resilient if and only if $\exists \mathcal{T}_0\ M_0^{a,b}(\mathcal{A}, \mathcal{T}_0)$ accepts.

Now suppose inductively that we have, for each $a' \in \mathbb{Z}^+$ and $b' \in \mathbb{N}$, algorithms $M_k^{a',b'}$ satisfying (Eq. 6) with $n = k$. Fix some $a \in \mathbb{Z}^+$ and $b \in \mathbb{N}$, and we define an algorithm $M_{k+1}^{a,b}$ which takes inputs of the form $(\mathcal{A}, \mathcal{T}, \mathcal{T}', \mathcal{T}_1, \ldots, \mathcal{T}_k, \rho, \rho_1, \ldots, \rho_k)$. Let $\mathcal{T} = (\tau, \sigma, j)$, $\mathcal{T}' = (\tau', \sigma', j')$, and $\rho = (r, \sigma^*, i)$. Furthermore, let t_0 denote the global time in the initial configuration $\mathcal{S}_0$, $|\tau|$ denote the length of τ, $\mathcal{S}'_{i+1}$ denote the initial configuration of τ', t_i denote the global time in the ith configuration $\mathcal{S}_i$ of τ, and $d_i = t_i - t_0$. We now describe the run of $M_{k+1}^{a,b}$ on this input.

First, check that τ and τ' are compliant traces to a goal configuration. Then, check if $d_i \leq a$; if this check fails, then we halt and accept, since by Definition 11, update rules cannot be applied after more than a time steps. Then, check if $\mathcal{S}_i \longrightarrow_r \mathcal{S}'_{i+1}$, by applying the ground substitution σ^* to r and checking that it is applicable to $\mathcal{S}_i$. If this checks fails, then we halt and accept, since r is not an applicable update rule to $\mathcal{S}_i$. Otherwise, check that $\mathcal{S}'_{i+1}$ is the correct result of applying this instance of r to $\mathcal{S}_i$. If this check fails, then reject, since τ' cannot be a valid reaction trace. Finally, let $A' = (\mathcal{R}, \mathcal{GS}, \mathcal{CS}, \mathcal{E}, \mathcal{S}'_{i+1})$, and simulate $M_k^{a-d_i,b}$ on the input $(A', \mathcal{T}', \mathcal{T}_1, \ldots, \mathcal{T}_k, \rho_1, \ldots, \rho_k)$. If the result of this simulation is that $M_k^{a-d_i,b}$ accepts the input, then we halt and accept, since by the inductive hypothesis, τ' must be a $(k, a - d_i, b)$-resilient reaction trace. Otherwise, we reject.

Taking into account the inductive hypothesis and Remark 4, it is clear that $M_{k+1}^{a,b}$ runs in polynomial time in the size of its input. Furthermore, it follows immediately by inspection of Definition 11 that A is $(k+1, a, b)$-resilient if and only if

$$\exists \mathcal{T}_0 \forall \rho_1 \exists \mathcal{T}_1 \ldots \forall \rho_{k+1} \exists \mathcal{T}_{k+1} \ M_{k+1}^{a,b}(A, \mathcal{T}, \mathcal{T}_1, \ldots, \mathcal{T}_{k+1}, \rho, \rho_1, \ldots, \rho_{k+1}) \text{ accepts.}$$

This concludes the inductive argument. It follows immediately from the quantifier-alternation characterization of PH that the (n, a, b)-resilience problem for η-simple PPSs with traces containing only facts of bounded size is in $\Sigma_{2n+1}^{\mathsf{P}}$. □

Remark 5. Even without assuming η-simplicity, a slight variation of the above argument gives a decision procedure of complexity $\Sigma_{2n+2}^{\mathsf{P}}$ for the (n, a, b)-resilience problem for PPSs with traces containing facts of bounded size. To modify the argument, we allow each universal quantifier to range over an additional ground substitution, which is used in the verification algorithm $M_n^{a,b}$ to check that an arbitrary configuration in the preceding witness trace is non-critical. Note that this check can be done in polynomial time (cf. Remark 4). If this check succeeds for *all* configurations and *all* such ground substitutions, then every witness trace is compliant.

In fact, even for 1-simple PPSs, the (n, a, b)-resilience problem is $\Sigma_{2n+1}^{\mathsf{P}}$-hard. We show this by a reduction from Σ_{2n+1}-SAT, the language of true quantified Boolean formulas (QBF) with $2n + 1$ quantifier alternations, where the first quantifier is existential and the underlying propositional formula is in 3-CNF form. This problem is known to be $\Sigma_{2n+1}^{\mathsf{P}}$-complete [38]. Recall that the truth of a quantified Boolean formula can be analyzed by considering the *QBF evaluation game* for the formula. In this game, two players, *Spoiler* and *Duplicator*, take turns choosing assignments to the formula's quantified variables. Duplicator chooses assignments for existentially-quantified variables with the goal of satisfying the underlying Boolean formula, while Spoiler chooses assignments for universally-quantified variables with the goal of falsifying it. The game concludes once assignments have been chosen for all of the quantified variables. A QBF ψ is true if and only if Duplicator has a winning strategy in this game [35].

In our reduction, we encode the positions of this QBF evaluation game into configurations, where a position of the QBF evaluation game for a formula

$$\psi := \exists \overline{v}_1 \forall \overline{v}_2 \exists \overline{v}_3 \ldots \forall \overline{v}_{2n} \exists \overline{v}_{2n+1} \varphi(\overline{v}_1, \overline{v}_2, \overline{v}_3, \ldots, \overline{v}_{2n+1})$$

is a sequence $\mathcal{P} = V_1, \ldots, V_j$ of assignments to the variables in $\overline{v}_1, \ldots, \overline{v}_j$ for some $j \leq 2n+1$. If j is even, then we say that the position $\mathcal{P}$ *belongs to Duplicator*; otherwise, we say that it *belongs to Spoiler*. The player who owns a given position makes the next move, choosing an assignment for the variables in the tuple $\overline{v}_{j+1}$. We use system rules to model assignments made by Duplicator, while update rules are used to model assignments made by Spoiler. Intuitively, the goal configurations are those positions of the game which encode assignments satisfying the underlying formula φ.

Theorem 2. *For all $a \in \mathbb{Z}^+$ and $b \in \mathbb{N}$, there exists a polynomial-time reduction from the Σ_{2n+1}-SAT problem to the (n, a, b)-resilience problem. Furthermore, the computed instance is always a 1-simple progressing planning scenario with traces containing only facts of bounded size.*

Proof. (sketch) Let $\psi := \exists \overline{v}_1 \forall \overline{v}_2 \exists \overline{v}_3 \ldots \forall \overline{v}_{2n} \exists \overline{v}_{2n+1} \varphi(\overline{v}_1, \overline{v}_2, \overline{v}_3, \ldots, \overline{v}_{2n+1})$ be an instance of Σ_{2n+1}-SAT, where the $\overline{v}_i$ are tuples of variables and φ is a 3-CNF formula. We can compute a 1-simple progressing planning scenario $A = (\mathcal{R}, \mathcal{GS}, \mathcal{CS}, \mathcal{E}, \mathcal{S}_0)$ which is (n, a, b)-resilient if and only if ψ is true. To do this, the initial configuration $\mathcal{S}_0$ contains 0-ary facts of the form Unk_i which indicate that the assignment to $\overline{v}_i$ is unknown, for each $1 \leq i \leq 2n+1$. We also include a 0-ary fact Rnd_0 indicating that no rounds of the QBF game have been played, and a 0-ary fact corresponding to each clause of φ. This represents the initial position of the QBF evaluation game.

We include system (resp. update) rules corresponding to assignments to the $\overline{v}_i$ tuples of variables for even (resp. odd); these rules consume the fact Unk_i and create facts of the form $Val_i(\overline{b})$, where $\overline{b}$ is a tuple Boolean values (*true* or *false*). These rules simulate moves of the QBF evaluation game, and change the configuration to represent the next position of the game. Furthermore, each rule of this kind can only be played when the appropriate Rnd_i fact is in the current configuration, and it increments the round counter from Rnd_i to Rnd_{i+1}. This ensures that the players can only choose assignments from positions that belong to them. We also include "verification" rules, which are used to check if the assignment after the conclusion of the game (encoded by the Val_i facts) satisfies φ. The goal configurations are those in which the final assignment has been verified to satisfy φ.

This encoding does not depend on the parameters a and b of the resilience problem: A admits an $(n, 1, 0)$-resilient trace if and only if it admits an (n, a, b)-resilient trace for all $a \in \mathbb{Z}^+$ and $b \in \mathbb{N}$. It follows easily from the simulation of the QBF evaluation game that, for all $a \in \mathbb{Z}^+$ and $b \in \mathbb{N}$, A is (n, a, b)-resilient if and only if Duplicator has a winning strategy for the QBF evaluation game for the formula ψ. □

A detailed specification of the reduction can be found in the technical report [3].

Corollary 1. *The (n,a,b)-resilience problem for η-simple PPSs with traces containing only facts of bounded size is Σ^P_{2n+1}-complete.*

5 Verifying Resilience in Maude

To experiment with resilience, we specified our running example of a travel planning scenario in the Maude rewriting logic language [11]. In contrast to the multiset rewriting representation, the Maude specification uses data structures, not facts, to represent system structure and state. The passing of time is modeled using rule duration. For example, the rule that models taking a flight takes time according to the duration of the flight. These rules combine an instantaneous rule with a time-passing rule. These design decisions, together with relegating as much as possible to equational reasoning, help reduce the search state space.

In the travel system scenarios, the goal is to attend a given set of events. System updates change flight schedules; goal updates either change event start time or duration, or add an event. Execution traces terminate when the last event is attended or an event is missed. Thus, for simplicity, we fix $a + b$ to be the end of the last possible event and leave it implicit. In the following, we describe the representation of key elements of the travel system specification: system state, execution rules, and updates. We then explain the algorithm for checking (n,a,b)-resilience and report on some simple experiments.

Representing Travel Status. The two main sorts in the travel system specification are Flight and Event. A term fl(cityD,cityA,fn,depT,dur) represents a flight, where cityD and cityA are the departure and arrival cities, fn is the flight number (a unique identifier), depT is the departure time, and dur is the duration. The departure time and duration are represented by hour-minute terms hm(h,m). For simplicity, flights are assumed to go at the same time every day, and all times are in GMT. A flight instance (sort FltInst) represents a flight on a specific date by a term: fi(flt,dtDep,dtArr) where flt is a flight, dtDep and dtArr are date-time terms representing the date and time of departure and arrival respectively. A date-time term has the form dt(yd,hm) where yd is a year-day term yd(y,d) and hm is an hour-minute term as above.

An event is represented by a term ev(eid,city,loc,yd,hm,dur,opt), where eid is a unique (string) identifier and the opt Boolean specifies if the event is optional; the other arguments are as above. A term of the form {conf} (sort Sys), where conf (sort Conf) is a multiset of configuration elements, represents a system execution state. The sorts of configuration elements are TConf, Log, and update descriptions. Terms of sort TConf represent a traveler's state, with one of three forms:

```
tc(dt,city,loc,evs) - planning
tc(dt,city,loc,evs,ev,fltil) - executing
tcCrit(dt,city,loc,evs,ev,reason) - critical
```

Here, `dt` is a date-time term, the travelers current date and time, and `city` tells what city the traveler is currently in. The term `loc` gives the location within the city, either the `airport` or the city `center`. The term `evs` is the set of events to be attended, while `ev` is the next event to consider. The term `fltil` is the flight instance list chosen to get from `city` to the location of `ev`. The constructor `tcCrit` signals a critical configuration in which a required event has been missed.

An update description is a term of the form `di(digs)` or `di(digs,n)` where `digs` is a set of digressions. Each digression describes an update to be applied by an update rule, and `n` bounds the number of updates that can be applied. Two kinds of updates are currently implemented: flight/system updates and event/goal updates. The flight updates are: `cancel`, which cancels the current flight, `delay(hm(h,m))`, which delays the current flight by `h` hours, `m` minutes, and `divert(city0,city1)`, which diverts the current flight from `city0` to `city1`, where the current flight is the first element of the flight instance list of an executing TConf term. The event/goal updates are: `edEvS(hm(h,m))`, which starts the current event `h` hours, `m` minutes earlier, `edEvD(hm(h,m))`, which extends the current event duration by `h` hours, `m` minutes, and `addEv(eid)`, which adds the event with id `evid` to the set of pending events.

Flight updates are only applied to the next flight the traveler is about to take. Similarly, the changes in event start time or duration are only applied to the next event to attend. This is a simplified setting, but sufficiently illustrates our formalism; more complex variations are possible. Lastly, an element of sort `Log` is a list of log items. It is used to record updates, flights taken, and events attended or missed. Among other things, when searching for flights, it is used to know which flights have been canceled.

Rewrite Rules. There are five system rules (`plan`, `noUFlts`, `flt`, `event`, and `replan`) and two update rules (`fltDigress` and `evDigress`, for flight/system updates and event/goal updates, respectively). The `plan` rule picks the event, `ev`, with the earliest start time from `nevs` and (non-deterministically) selects a list of flight instances, `fltil`, from the set of flight instance lists arriving at the event city before the start time. The set of possible flights is stored in a constant `FltDB`. `log1` is `log` with an item recording the rule firing added. The conditional if ... does the above computing.

```
crl [plan]: {tc(dt,city, loc, nevs) log}
          => {tc(dt, city, loc, evs0, ev, fltil) log1} if ...
```

The rule `noUFlts` handles cases in which there is no usable flight instance list given the traveler time and location and the time and location of the next event. If `ev` is optional then it is dropped (recording this in the log) and the rule `plan` is applied to the remaining events; otherwise, the configuration becomes critical and execution terminates. The rule `flt` models taking the next flight, assuming the flight departure time is later than the traveler's current time. This rule updates the traveler's time to the flight arrival time dtArr and traveler's city to the destination city1. Then, the flight instance taken is removed from the list.

```
crl [flt]: {tc(dt, city0, airport, evs,ev,flti ; fltil) conf}
=> {tc(dtArr, city1, airport, evs,ev, fltil) conf1} if ....
```

The event rule (not shown) models attending the currently selected event. It can be applied when the traveler city is the same as the event city and the current time is not after the event start. As for the flight rule, the current time is updated to the event end time and the traveler returns to the airport. If the traveler arrives at the event city too late, as for the noUFlts rule, if the event is optional, the event rule drops the event and enters a log item, otherwise it produces a critical TConf. The replan rule handles the situation in which the traveler city is not the event city and the next flight, if any, does not depart from the traveler city or has been missed. The pending flight instance list is dropped, the event is put back in the event set, and a log item is added to the log.

The update rule fltDigress only applies if the digression counter is greater than zero and the rule decrements the counter. A flight digression, fdig, is non-deterministically selected from the available digressions (the first argument to di).

```
crl [fltDigress]: {tconf di(fdig digs, s n) conf}
=> {tconf1 di(digs fdig,n) conf1}
    if tc(dt, city0, airport, evs,ev, flti ; fltil) := tconf
    ∧ city0 =/= getCity(ev)
    ∧ tconf1 conf1 := applyDigression(tconf,conf,fdig)
```

The first condition exposes the structure of tconf to ensure there is a pending flight to update. The auxiliary function applyDigression specifies the result of the update. For example, the cancel update removes flti from the list and adds a log item recording that this flight instance is cancelled. The case where the update description has the form di(fdig digs) is similar, except here fdig is removed when applied, and updating stops when there are no more update elements in the set. Similarly, the rule evDigress non-deterministically selects an event digression from the configuration's digression set and applies the auxiliary function applyEvDigress to determine the effect of the update. It only applies if the update counter is greater than zero, and the TConf component has a selected next event.

A planning scenario is defined by an initial system configuration iSys, a database of flights FltDB, a database of events EvDB, and an update description. A trace iSys -TR-> xSys is a sequence of applications of rule instances from the travel rules TR, leading from iSys to xSys. It is a *compliant goal trace* if xSys satisfies the goal condition (goal) that the traveler configuration has no remaining events, and no required events have been missed. Formally, the traveler component of xSys must have the form tc(dt,city,loc,mtE), since, when a required event is missed, it is rewritten to a term of the form tcCrit(dt,city,loc,evs,ev,comment).

Checking (n, a, b)-resilience. Checking (n, a, b)-resilience in the travel planning system is implemented by the equationally-defined function isAbRes using Maude's reflection capability and strategy language. As previously mentioned, the upper bound on time is implicitly determined by the times and durations of available events, not treated as a parameter.

The function isAbRes checks (n, a, b)-resilience by first using metaSearch to find a candidate goal state, then metaSearchPath gives the corresponding compliant goal trace.[2] The candidate trace is converted to a rewrite strategy (representing the trace's list of rule instances). The function checkAbRes iterates through the initial prefixes of the strategy, using metaSRewrite to follow the trace prefix. This implements a check for reaction traces at all possible points of update rule application (cf. Definition 11). For each state resulting from executing a prefix, the function checkDigs is called to apply each one of the available updates, and then we invoke isAbRes to check for an $(n-1, a-d, b)$-resilient extension trace, where d is the number of time steps up to the end of the prefix. If n is zero, abResCheck simply finds a compliant goal trace. If no $(n-1, a-d, b)$-resilient extension trace can be found, then the current candidate trace is rejected, and isAbRes continues searching for the next candidate trace. If the search for candidate traces fails, then the system under consideration is not (n, a, b)-resilient. If the check for an $(n-1, a-d, b)$-resilient extension trace succeeds for every update of every prefix execution, then the strategy is returned as a witness for (n, a, b)-resilience. We tested (n, a, b)-resilience to flight/system updates and event/goal updates with instances of the following command.

```
red isAbRes(['TRAVEL-SCENARIO],N,allDi,SYST,patT,tCond,uStrat,0).
red isAbRes(['TRAVEL-SCENARIO],1,allEv,iSysT,patT,tCond,ueStrat,0).
```

The results are summarized in Table 2. Note that N is the number of updates (1, 2, or 3), SYST is (the meta representation of) an initial state with a starting day that is as late as possible to succeed if nothing goes wrong (247) or one day earlier (246) and 2 or 3 events. patT and tCond are the metaSearch pattern and condition arguments and uStrat is used to construct the update rule strategy. In the summary tables the R? indicates the result of isAbRes: Y for yes (a non-empty strategy is returned) and N for no. A dash indicates the experiment not done (because the check fails for smaller N). Lastly, allEv is the set of all implemented event update descriptions.

---

[2] Execution terminates if a critical state is reached, so paths to goal states are always compliant.

N:	1		2		3	
2ev	R?	time	R?	time	R?	time
247	N	86ms	-	-	-	-
246	Y	81ms	Y	147ms	N	7476ms
3ev	R?	time	R?	time	R?	time
247	N	1400ms	-	-	-	-
246	Y	325ms	Y	685ms	NF	-

(a) flight/system update rules

N:	1		2		3	
2ev	R?	time	R?	time	R?	time
247	Y	78ms	N	77ms	-	-
246	Y	98ms	N	34800ms	-	-
3ev	R?	time	R?	time	R?	time
247	Y	143ms	N	2627ms	-	-
246	Y	220ms	Y	633ms	Y	2634ms

(b) event/goal update rules

Fig. 2. Summary of (n, a, b)-resilience experiments

6 Conclusions and Related Work

We have shown that, for η-simple PPSs with traces containing only facts of bounded size, the (n, a, b)-resilience problem is Σ_{2n+1}^{P}-complete. In [1], we showed that the version of this problem without time bounds is PSPACE-complete for balanced systems with traces containing only facts of bounded size. In addition to the formal model and complexity results, we have implemented automated verification of time-bounded resilience using Maude. Resilience has been studied in diverse areas such as civil engineering [8], disaster studies [29], and environmental science [16]. Formalizations of resilience are often tailored to specific applications [4,19,27,30,37] and cannot be easily adapted to different systems. However, while we illustrated time-bounded resilience via an example of a PPS modeling a flight planning scenario, our earlier work has studied similar properties for a diverse range of other critical, time-sensitive systems, from collaborative systems subject to governmental regulations [24], to distributed unmanned aerial vehicles (UAV) performing safety-critical tasks [21,22]. The strength of our formalism is its flexibility in modeling a wide range of multi-agent systems.

Interest in resilience and related concepts such as robustness [9], recoverability [22,33], fault tolerance [25], and reliability [5], has grown in recent years. In [31,40], the authors define a notion of robustness in which a system is robust if the actions taken by an adversary cannot force the system to release more information than it would in the absence of the adversary's actions. In [33], the authors study *time-bounded recovery* for *logical scenarios*, which are families of system states represented by patterns whose variables are constrained to describe an operating domain, and where recoverability is parameterized by an ordered set of safety conditions. Intuitively, a t-recoverable logical scenario is one which, when operating in normal mode, can recover from a lower-level safety condition to an optimal safety condition within t time steps, without reaching an unsafe condition. Like our formalization, t-recoverability concerns recovery from a deviation from normal execution. The primary distinction in [33] is that deviations are internal to the system, rules update the model state and compute control commands, and enabled rules must fire before time passes, as is common in real-time systems.

Our definition of time-bounded resilience (Definition 11) can be seen as a modification of [1, Definitions 9–10], with time parameters a and b and taking into account both system and goal updates. Here, the recoverability conditions from [1] are simplified to the total relation on configurations. Further investigation of recoverability conditions and resilience with respect to update rules that consume or create critical facts is left for future work. Another avenue of future work is to find conditions beyond η-simplicity which allow for polynomial-time solvability of the trace compliance problem. We also plan to study time-bounded resilience problems with respect to update rules that consume or create critical facts, and other variations involving, *e.g.*, real-time models and infinite traces. We are interested in the relationship between resilience and other properties of time-sensitive distributed systems [22], such realizability, recoverability, reliability, and survivability, as well as in specific applications of resilience, where further implementation results could provide interesting insights.

Acknowledgment. We thank Vivek Nigam for many insightful discussions during the early part of this work. Kanovich was partially supported by EPSRC Programme Grant EP/R006865/1: "Interface Reasoning for Interacting Systems (IRIS)". Scedrov was partially supported by the U. S. Office of Naval Research under award number N00014-20-1-2635 during the early part of this work. Talcott was partially supported by the U. S. Office of Naval Research under award numbers N00014-15-1-2202 and N00014-20-1-2644, and NRL grant N0017317-1-G002.

References

1. Alturki, M.A., Ban Kirigin, T., Kanovich, M., Nigam, V., Scedrov, A., Talcott, C.: On the formalization and computational complexity of resilience problems for cyber-physical systems. In: Theoretical Aspects of Computing–ICTAC 2022: 19th International Colloquium, Tbilisi, Georgia, September 27–29, 2022, Proceedings, pp. 96–113. Springer, Berlin (2022)
2. Arora, S., Barak, B.: Complexity Theory: A Modern Approach. Cambridge University Press Cambridge (2009)
3. Ban Kirigin, T., Comer, J., Kanovich, M., Scedrov, A., Talcott, C.: Technical report: Time-bounded resilience (2024). arXiv:2401.05585
4. Banescu, S., Ochoa, M., Pretschner, A.: A framework for measuring software obfuscation resilience against automated attacks. In: 2015 IEEE/ACM 1st International Workshop on Software Protection, pp. 45–51 (2015)
5. Bauer, E.: Design for Reliability: Information and Computer-Based Systems. Wiley, New York (2011)
6. Bennaceur, A., Ghezzi, C., Tei, K., Kehrer, T., Weyns, D., Calinescu, R., Dustdar, S., Hu, Z., Honiden, S., Ishikawa, F., Jin, Z., Kramer, J., Litoiu, M., Loreti, M., Moreno, G., Müller, H., Nenzi, L., Nuseibeh, B., Pasquale, L., Reisig, W., Schmidt, H., Tsigkanos, C., Zhao, H.: Modelling and analysing resilient cyber-physical systems. In: 2019 IEEE/ACM 14th International Symposium on Software Engineering for Adaptive and Self-Managing Systems (SEAMS), pp. 70–76 (2019)

7. Bloomfield, R., Fletcher, G., Khlaaf, H., Ryan, P., Kinoshita, S., Kinoshit, Y., Takeyama, M., Matsubara, Y., Popov, P., Imai, K., et al.: Towards identifying and closing gaps in assurance of autonomous road vehicles–a collection of technical notes part 1 (2020). arXiv:2003.00789
8. Bozza, A., Asprone, D., Fabbrocino, F.: Urban resilience: A civil engineering perspective. Sustainability **9**(1) (2017)
9. Bruneau, M., Chang, S.E., Eguchi, R.T., Lee, G.C., O'Rourke, T.D., Reinhorn, A.M., Shinozuka, M., Tierney, K., Wallace, W.A., Von Winterfeldt, D.: A framework to quantitatively assess and enhance the seismic resilience of communities. Earthq. Spectra **19**(4), 733–752 (2003)
10. Caminiti, S., Finocchi, I., Fusco, E.G., Silvestri, F.: Resilient dynamic programming. Algorithmica **77**(2), 389–425 (2017)
11. Clavel, M., Durán, F., Eker, S., Lincoln, P., Martí-Oliet, N., Meseguer, J., Talcott, C.: All About Maude: A High-Performance Logical Framework, volume 4350 of LNCS. Springer, Berlin (2007)
12. Cunningham, D., Grove, D., Herta, B., Iyengar, A., Kawachiya, K., Murata, H., Saraswat, V., Takeuchi, M., Tardieu, O.: Resilient x10: Efficient failure-aware programming. SIGPLAN Not. **49**(8), 67–80 (2014)
13. Durgin, N.A., Lincoln, P., Mitchell, J.C., Scedrov, A.: Multiset rewriting and the complexity of bounded security protocols. J. Comput. Secur. **12**(2), 247–311 (2004)
14. Eigner, O., Eresheim, S., Kieseberg, P., Klausner, L.D., Pirker, M., Priebe, T., Tjoa, S., Marulli, F., Mercaldo, F.: Towards resilient artificial intelligence: Survey and research issues. In: 2021 IEEE International Conference on Cyber Security and Resilience (CSR), pp. 536–542 (2021)
15. Ferraro-Petrillo, U., Finocchi, I., Italiano, G.F.: Experimental study of resilient algorithms and data structures. In: Festa, P., (ed.) Experimental Algorithms, pp. 1–12. Springer, Berlin (2010)
16. Folke, C.: Resilience: the emergence of a perspective for social-ecological systems analyses. Global Environ. Change **16**(3):253–267 (2006). Resilience, Vulnerability, and Adaptation: A Cross-Cutting Theme of the International Human Dimensions Programme on Global Environmental Change
17. Goel, S., Hanneke, S., Moran, S., Shetty, A.: Adversarial resilience in sequential prediction via abstention. Adv. Neural Inf. Process. Syst. **36** (2024)
18. Hirshfeld, Y., Rabinovich, A.: Logics for real time: decidability and complexity. Fund. Inform. **62**(1), 1–28 (2004)
19. Huang, W., Zhou, Y., Sun, Y., Banks, A., Meng, J., Sharp, J., Maskell, S., Huang, X.: Formal verification of robustness and resilience of learning-enabled state estimation systems for robotics (2020)
20. Hukerikar, S., Diniz, P.C., Lucas, R.F.: A programming model for resilience in extreme scale computing. In: IEEE/IFIP International Conference on Dependable Systems and Networks Workshops (DSN 2012), pp. 1–6 (2012)
21. Kanovich, M., Ban Kirigin, T., Nigam, V., Scedrov, A., Talcott, C.: Timed multiset rewriting and the verification of time-sensitive distributed systems. In: 14th International Conference on Formal Modeling and Analysis of Timed Systems (FORMATS) (2016)
22. Kanovich, M., Ban Kirigin, T., Nigam, V., Scedrov, A., Talcott, C.: On the complexity of verification of time-sensitive distributed systems. In: Dougherty, D., Meseguer, J., Mödersheim, S.A., Rowe, P., (eds.), Protocols, Strands, and Logic, volume 13066 of Springer LNCS, pp. 251–275. Springer International Publishing (2021)

23. Kanovich, M., Ban Kirigin, T., Nigam, V., Scedrov, A., Talcott, C.L.: Time, computational complexity, and probability in the analysis of distance-bounding protocols. J. Comput. Secur. **25**(6), 585–630 (2017)
24. Kanovich, M., Ban Kirigin, T., Nigam, V., Scedrov, A., Talcott, C.L., Perovic, R.: A rewriting framework and logic for activities subject to regulations. Math. Struct. Comput. Sci. **27**(3), 332–375 (2017)
25. Koren, I., Krishna, C.M.: Fault-Tolerant Systems. Morgan Kaufmann (2020)
26. Koutsoukos, X., Karsai, G., Laszka, A., Neema, H., Potteiger, B., Volgyesi, P., Vorobeychik, Y., Sztipanovits, J.: Sure: a modeling and simulation integration platform for evaluation of secure and resilient cyber-physical systems. Proc. IEEE **106**(1), 93–112 (2018)
27. Madni, A.M., Erwin, D., Sievers, M.: Constructing models for systems resilience: challenges, concepts, and formal methods. Systems **8**(1) (2020)
28. Madni, A.M., Jackson, S.: Towards a conceptual framework for resilience engineering. IEEE Syst. J. **3**(2), 181–191 (2009)
29. Manyena, S.B.: The concept of resilience revisited. Disasters **30**(4), 434–450 (2006)
30. Mouelhi, S., Laarouchi, M.-E., Cancila, D., Chaouchi, H.: Predictive formal analysis of resilience in cyber-physical systems. IEEE Access **7**, 33741–33758 (2019)
31. Myers, A.C., Sabelfeld, A., Zdancewic, S.: Enforcing robust declassification and qualified robustness. J. Comput. Secur. **14**(2), 157–196 (2006)
32. Neches, R., Madni, A.M.: Towards affordably adaptable and effective systems. Syst. Eng. **16**(2), 224–234 (2013)
33. Nigam, V., Talcott, C.L.: Automating recoverability proofs for cyber-physical systems with runtime assurance architectures. In: David, C., Sun, M., (eds.) 17th International Symposium on Theoretical Aspects of Software Engineering, volume 13931 of Lecture Notes in Computer Science, pp. 1–19. Springer, Berlin (2023)
34. Olowononi, F.O., Rawat, D.B., Liu, C.: Resilient machine learning for networked cyber physical systems: a survey for machine learning security to securing machine learning for cps. IEEE Commun. Surv. Tutorials **23**(1), 524–552 (2021)
35. Papadimitriou, C.H.: Computational Complexity. Academic Internet Publication (2007)
36. Prasad, A.: *Towards Robust and Resilient Machine Learning*. Ph.D. thesis, Carnegie Mellon University (2022)
37. Sharma, V.C., Haran, A., Rakamaric, Z., Gopalakrishnan, G.: Towards formal approaches to system resilience. In: 2013 IEEE 19th Pacific Rim International Symposium on Dependable Computing, pp. 41–50 (2013)
38. Stockmeyer, L.J.: The polynomial-time hierarchy. Theor. Comput. Sci. **3**(1), 1–22 (1976)
39. Vardi, M.: Efficiency versus resilience: What covid-19 teaches computing. Commun. ACM **63**(5), 9 (2020)
40. Zdancewic, S., Myers, A.C.: Robust declassification. In: Proceedings of the 14th IEEE Workshop on Computer Security Foundations, CSFW '01, p. 5. IEEE Computer Society, USA (2001)

Verifying Safe Memory Reclamation in Concurrent Programs with CafeOBJ

Duong Dinh Tran[✉][image] and Kazuhiro Ogata[image]

Japan Advanced Institute of Science and Technology, Ishikawa 923-1211, Japan
duongtd@jaist.ac.jp, ogata@jaist.ac.jp

Abstract. Verification of concurrent programs is very tough because the proof needs to account for a huge number of interleavings of execution steps of all threads. As concurrent programs are often implemented in C/C++, which do not support garbage collection, program verification needs to consider memory reclamation as well. In this paper, we report the safe memory verification of two concurrent programs integrated with hazard pointers, a mechanism for safe memory reclamation. We use CafeOBJ to formally specify the two programs and the properties of interest. The formal proofs are produced with the assistance of a tool called IPSG.

Keywords: Formal verification · Memory reclamation · CafeOBJ

1 Introduction

Over the years, concurrent programs have been widespread, in part because they allow to speed up the execution performance when they are executed in parallel, leveraging multicore hardware. The concurrency, however, makes it notoriously difficult to implement a concurrent program and to verify it acts as expected, i.e., all the desired properties are satisfied and there are no undesirable ones (bugs). This is mainly due to a huge number of interleavings happening between the execution steps of all threads.

The majority of concurrent programs and concurrent data structures are implemented in unsupported garbage collection languages, like C and C++. That means programmers are responsible for deallocating unused memory. However, unlike sequential programs, it is not simple to deallocate a memory block that is no longer in use by a thread in concurrent programs since that block of memory may still be accessed by other threads. This raises one more challenge in reasoning about concurrent programs: *safety of memory reclamation*, that is, a no-longer-used memory block by a thread should be safely deallocated with a guarantee that no other thread accesses such a freed memory block after that.

Some mechanisms to provide safe memory reclamation have been proposed, such as hazard pointers [14], read-copy-update [13], and epoch-based reclamation [7]. With the hazard pointers-based solution, each thread additionally has

---

This work was supported by JSPS KAKENHI Grant Number 23K28060.

© The Author(s), under exclusive license to Springer Nature Switzerland AG 2024
K. Ogata and N. Martí-Oliet (Eds.): WRLA 2024, LNCS 14593, pp. 45–61, 2024.
https://doi.org/10.1007/978-3-031-65941-6_3

its own global pointer, which points to the shared object address it is accessing. A thread can read but cannot modify the hazard pointers of other threads. Before deallocating a memory block, a thread needs to check that the hazard pointers of all other threads do not point to this block.

In this paper, we formally verify the safe memory reclamation of two concurrent programs, the Shared counter and Treiber's stack [24], integrated with the hazard pointer mechanism. We formally specify the two programs in CafeOBJ [5], an algebraic specification language that supports writing formal specifications of a wide variety of systems. CafeOBJ can be also used as an interactive theorem proving system, where verifiers are supposed to write the so-called *proof scores* [17] and execute them with CafeOBJ. We use a tool called IPSG [23], which can automate the manual writing task, to produce the proof scores. Precisely, given a CafeOBJ formal specification, invariant properties formalizing the desired properties we want to verify, and an auxiliary lemma list, IPSG can automatically produce the proof scores verifying those properties. Note that human users are in charge of conjecturing the auxiliary lemmas. There are some existing studies on the verification of the two programs [9,10,18,21], which will be discussed in Sect. 6. The strong point of our verification is that the proofs are based on induction, a standard and pedagogical proof technique, so that readers easily grasp the basic idea even for those unfamiliar with formal verification. We use CafeOBJ for both specification and verification tasks so that the entire verification process can be carried out smoothly. Although we need to construct many auxiliary lemmas to complete the verification, we have learned that the lemmas can be divided into some groups as discussed in Sect. 4.3, based on which a new lemma can be conjectured. We regard this as an initial step towards verifying concurrent programs integrated with some advanced safe memory reclamation mechanisms with CafeOBJ more efficiently.

We make the CafeOBJ formal specifications, proof scores, and other verification materials available on the webpage[1]. In the rest of this section, we elaborate on the Shared counter program and its fixed version with hazard pointers. Then, Sect. 2 provides some necessary preliminaries. Afterward, Sects. 3 and 4 describe how to formally specify the Shared counter program in CafeOBJ and verification of the safe memory reclamation, respectively. The verification result of Treiber's concurrent stack is briefly reported in Sect. 5. Some closely related work is mentioned in Sect. 6, and summarization of our paper is recapitulated in Sect. 7.

1.1 Shared Counter Program

The program is shown in Fig. 1a. Multiple threads share a global counter c. Each thread can perform the increment function inc(), which returns the current value of c and increments it. To do so, the thread first allocates a new memory block n (line l1), where the updated counter will be stored after the increment succeeds. The thread then reads the value of c into its local pointer p (line l2 and Fig. 1b), gets the value v of the counter (line l3), and stores the successor of v into n

---

[1] https://github.com/duongtd23/Reclamation-CafeOBJ-verification.

(line l4). The symbol CAS in line l5 denotes the atomic primitive compare-and-swap, which takes a pointer and two values. It dereferences the pointer and compares the dereferenced value with the first given value. If they are the same, it updates the pointer with the second given value and returns true. Otherwise, false is just returned. In the CAS in line l5 of Fig. 1a, because &C is passed as the first parameter, the dereferenced value, i.e., *(&C), of this parameter is simplified to C. Therefore, this CAS updates C to n, the address that stores the updated value of the counter, if C and p point to the same address (Fig. 1c); otherwise, the thread reads the value of C into p again and repeats the procedure. Figure 1b graphically visualizes the values of p and n of thread T_1 and C right after T_1 reads the value of C, i.e., 0xf1, into p. In this state, n is pointing to the address 0xf2. Figure 1c shows the state right after T_1 succeeds the CAS, in which C is now pointing to the same address as n, where stores the value 2.

```
     int *C = new int(0);
l0   fun inc() {
l1'      int v, *p, *n;
l1       n = new int;
l2'      do {
l2           p = C;
l3           v = *p;
l4           *n = v+1;
l5       } while(!CAS(&C,p,n));
l6       free(p); // errors occur
l7       return v;
l7'  }
```

(a) Shared counter.

(b) Right after T_1 reads the value of C into its local p.

(c) Right after T_1 successfully completes the CAS.

Fig. 1. If the statement at line l6 is disabled, a memory leak will happen. If that statement is enabled, other errors will occur because of the concurrency

After a successful increment, the thread needs to deallocate the memory block previously occupied by C (line l6); otherwise, a memory leak will happen. In the illustration visualized in Fig. 1c, that memory block is the one with address 0xf1. However, if the statement free(p) is enabled, an error will occur in the following scenario. Thread T_1 invokes inc(), successfully loading C into its p, but has not yet attempted to read the value pointed by p. In other words, T_1 completes its execution at l2 and stops at l3 (we simply say that T_1 is located at l3). Another thread T_2 also performs inc(), successfully completes the CAS and frees the memory pointed by its local pointer p, i.e., the old C. Now T_1 tries to read the value pointed by its p at line l3, and causes an error due to the address pointed by p is no longer available as it was deallocated by T_2.

```
        int *C = new int(0);                   Set detached[N+1] = {∅};
        int *hp[N+1] = {0};               f0  fun reclaim(int *p) {
    l0  fun inc() {                       f1    insert(detached[tid],p);
    l1'   int v, *p, *n;                  f2    Set inuse = empty;
    l1    n = new int;                    f3    while(!isEmpty(detached[tid])) {
    l2'   do {                            f4      bool isfree = true;
    l2      hp[tid] = C;                  f5      int *m = pop(detached[tid]);
    l3      p = hp[tid];                  f6      for(i=1; i<=N && isfree; i++) {
    l4      v = *p;                       f6        if (hp[i] == m && i != tid)
    l5      *n = v+1;                     f6          isfree = false; }
    l6    } while(!CAS(&C,p,n));          f7      if (isfree) free(m);
    l7    reclaim(p); hp[tid] = null;     f7      else insert(inuse, m);
    l8    return v;                       f7'   }
    l8' }                                 f8    moveAll(detached[tid], inuse);
                                          f8' }
```

Fig. 2. Shared counter—A safe memory reclamation version with hazard pointers

Figure 2 shows the fixed version of the shared counter program based on hazard pointers [14]. We suppose that there are N threads involved in this program, whose identifiers (IDs) range from 1 to N. Hazard pointers of the N threads are declared as the global pointer array hp (note that hp is declared to be of size N+1 because the threads have identifiers from 1 to N). At line l2, the thread sets its hazard pointer to the shared counter C, where tid denotes its ID ($1 \leq$ tid $\leq$ N). Then, it assigns the hazard pointer to its local pointer p (line l3). At line l7, instead of immediately deallocating the object p, the thread invokes function reclaim(p). In this function, p is first inserted into its detached set (line f1). For each object m in this set, the thread scans the hazard pointers of all other threads to check that all are different from m. If so, m is safe to deallocate, otherwise, it is inserted into the set detached again to be reclaimed in the next invocation.

2 Preliminaries

This section first gives the definitions of Observational Transition System, which will be used to formalize the programs later. We then introduce CafeOBJ in a nutshell.

2.1 Observational Transition System

Observational Transition System (OTS) [17] is a kind of state machine used to formalize systems. Let Υ denote a universal state space. Let D_o, D_t, D_o^i, and D_t^i be sets of data values, where i serves as indices.

Definition 1. An OTS $\mathcal{S}$ is a tuple $\langle \mathcal{O}, \mathcal{I}, \mathcal{T} \rangle$ in which:

- $\mathcal{O}$: A finite set of *observation functions* (or observers). Each observation function $o : \Upsilon \times D_o^1 \times \ldots \times D_o^m \to D_o$ takes one state and m ($m \geq 0$) data values and returns one data value. The equivalence relation ($v_1 =_\mathcal{S} v_2$) between two states v_1 and v_2 is defined as $(\forall o \in \mathcal{O}), (\forall x_1 \in D_o^1), \ldots, (\forall x_m \in D_o^m). (o(v_1, x_1, \ldots, x_m) = o(v_2, x_1, \ldots, x_m))$.
- $\mathcal{I}$: The set of *initial states*, where $\mathcal{I} \subseteq \Upsilon$.
- $\mathcal{T}$: A finite set of *transitions*. Each transition $t : \Upsilon \times D_t^1 \times \ldots \times D_t^n \to \Upsilon$ takes one state and n ($n \geq 0$) data values, and returns one state. Each transition t has the *effective condition* $c\text{-}t : \Upsilon \times D_t^1 \times \ldots \times D_t^n \to \text{Bool}$. If $c\text{-}t(v, x_1, \ldots, x_n)$ does not hold, then $t(v, x_1, \ldots, x_n) =_\mathcal{S} v$ for $x_1 \in D_t^1, \ldots, x_n \in D_t^n$.

Given an OTS $\mathcal{S}$, the set of all reachable states $\mathcal{R}_\mathcal{S}$ with respect to (w.r.t.) $\mathcal{S}$ is defined as follows. Each $v \in \mathcal{I}$ is reachable w.r.t. $\mathcal{S}$; and $t(v, x_1, \ldots, x_n)$ is reachable w.r.t. $\mathcal{S}$ if $v \in \Upsilon$ is reachable w.r.t. $\mathcal{S}$ for some $t \in \mathcal{T}$ and $x_i \in D_t^i$ for $i = 1, \ldots, n$. A predicate $p : \Upsilon \times D_1 \times \ldots \times D_l \to \text{Bool}$ is called a *state predicate*. A state predicate that holds in all reachable states is called an *invariant*.

2.2 CafeOBJ in a Nutshell

CafeOBJ [5] is a programming language equipped with a rich syntax for writing formal specifications of systems and protocols. Here we provide a brief introduction to CafeOBJ in the context of how to use it to formally specify the Shared counter program. The most basic building blocks of CafeOBJ are *modules*. A CafeOBJ module starts with the keyword `module` (abbreviated as `mod`, and possibly affixed with `*` or `!`, saying that the module has loose semantics or tight semantics [1], respectively) and may have the following declarations:

- importations of previously defined modules.
- sort declarations: `[S1 S2 ...]`, where `S1`, `S2`, ... are sort names.
- declarations of ordering sorts relation: `[S1 < S2]` denotes that `S1` is a subsort of `S2`.
- operator declarations: `op f : S1 ... Sn -> S {AT1 ... ATk}`, which declares an operator (or a function symbol) with n arguments of sorts `S1`, ..., `Sn` that produces an output of sort `S`. The optionally equational theory attributes `AT1`, ..., `ATk` may be, for example, `assoc` (associativity) and `comm` (commutativity). An operator without input arguments is called a constant.
- variable declarations: `vars V1 V2 ... : S`, where `V1`, `V2`, ... are variable names and `S` is a sort name.
- unconditional equations: `eq T1 = T2 .`, where `T1` and `T2` are terms.
- conditional equations: `ceq T1 = T2 if C .`, where `C` is a Boolean term.

For instance, to specify the "execution labels" `l1`, ..., `l8` and `f1`, ..., `f8` assigned to the functions `inc()` and `reclaim()` in Fig. 2, the following module is introduced:

```
mod! LABEL {
  [LabelI LabelR < Label]
  ops l1 l2 l3 l4 l5 l6 l7 l8 : -> LabelI {constr} .   -- for function inc()
  ops f1 f2 f3 f4 f5 f6 f7 f8 : -> LabelR {constr} .   -- for function reclaim()
  var I : LabelI          var R : LabelR
  eq (I = R) = false .
  ...      -- omitted some other equations
}
```

where some other equations are omitted by the symbol .... The attribute constr denotes that the operator is a constructor of its respective sort.

CafeOBJ provides the user with the open-close environment in the following form:

open MOD .
  ...
close

This syntax creates a temporary copy of the existing module MOD and adds new information into it by introducing new operators, equations, etc. at the placeholder .... This open-close environment is specifically useful when doing formal verification [17].

3 CafeOBJ Formal Specification of Shared Counter

We present the CafeOBJ formal specification of the program shown in Fig. 2 in this section.

Integers. We suppose that integers used in the program are unsigned. We define the set of unsigned integers as the union of Zero, which contains only 0, and NzUInt, which contains s(0) (i.e., 1), s(s(0)) (i.e., 2), and so on:

```
[Zero NzUInt < UInt]
op 0 :             -> Zero   {constr}
op s : UInt        -> NzUInt {constr}
op _+_ : UInt UInt -> UInt   {assoc comm prec: 33}
vars I I2 : UInt
eq (0 = s(I)) = false .
eq (s(I) = s(I2)) = (I = I2) .
eq I + 0 = I .
eq I + s(I2) = s(I + I2) .
```

In the specification of integers, we also define two predicates: less than < and less than or equal to <=.

The program execution is formalized as an OTS. Transitions of the OTS are the union of the transitions formalizing the function inc() and those formalizing the function reclaim(). Each transition specifies how a thread advances one single step of execution, such as a thread T assigns the value pointed by its local pointer p to its v. The combination of all transitions and all threads executing the transitions reflects the concurrent behaviors. Even if the number of threads is fixed to N, our specification lets N be an arbitrary number as it is but does not set it a concrete value.

The sort UInt is used to represent thread IDs as each thread ID is zero or a natural number. We introduce the sort Sys (standing for system), representing the state space. To keep track of the execution progress of each thread, we introduce an observation function, namely pc, with the following signature:

```
[Sys]
--         state    thread ID   execution label
op pc : Sys      UInt       -> Label
```

The observation function pc takes a state and a thread ID as inputs and returns the "execution label" where the thread is located at that state. The effective condition of a transition (i.e., the condition for triggering the transition proceeds) is usually specified using this observation function. For example, the transition specifying the assignment of the value pointed by thread T's pointer p to its v mentioned above cannot be triggered unless pc(S,T) is l4, where S denotes a state.

3.1 Function Inc

The shared counter and the thread-local variables are formalized as observation functions. Let S and T be an arbitrary state and a thread ID (i.e., an integer), respectively. We elaborate on these observation functions in terms of what inputs they take and what outputs they return as follows:

- counter(S): outputs the shared counter C at state S.
- v(S,T): outputs the value of v of thread T at state S, which is a valid integer or null.
- p(S,T): outputs the value of p of thread T at state S, which is a valid address or nil - a null pointer.
- n(S,T): outputs the value of n of thread T at state S.
- hp(S,T): outputs the hazard pointer of thread T at state S.
- *(S,XN): takes a state S and a pointer XN (possibly null) as inputs and outputs the value pointed by XN at state S.

Recall that, there is also the observation function pc, where pc(S,T) outputs the "execution label" where thread T is located at state S.

We use a constant—init—to denote arbitrary initial states. The initial states are defined in terms of equations for the observation functions as follows:

```
op init :      -> Sys {constr}
eq pc(init,T) = l1 .
eq counter(init) = c .
eq *(init,XN) = (if XN = c then 0 else null fi) .
eq v(init,T)  = null .
eq p(init,T)  = nil .
eq n(init,T)  = nil .
eq hp(init,T) = nil .
```

Initially, the shared counter is c, which can be arbitrary. Thus, *(init,X) will be 0 (suppose that initially, the value of the counter is 0) if the address X is c. All other addresses different from c have the value null initially. For each thread T, its v, p, and n all are null (which is denoted by nil in our CafeOBJ specification).

As mentioned before, each single step of execution in function inc() is specified by a transition. As an example, we show the definition of the so-called transition assign-v, which specifies the assignment of *p to v:

```
op assign-v : Sys UInt -> Sys {constr} .
eq counter(assign-v(S,T)) = counter(S) .
ceq pc(assign-v(S,T), T2) = (if T = T2 then l5 else pc(S,T2) fi)
  if c-assign-v(S,T) .
ceq v(assign-v(S,T), T2)  = (if T = T2 then *(S,p(S,T)) else v(S,T2) fi)
  if c-assign-v(S,T) .
eq p(assign-v(S,T), T2)   = p(S,T2) .
eq n(assign-v(S,T), T2)   = n(S,T2) .
eq hp(assign-v(S,T), T2)  = hp(S,T2) .
eq *(assign-v(S,T), XN)   = *(S,XN) .
ceq assign-v(S,T)         = S if not c-assign-v(S,T) .
eq c-assign-v(S,T)        = (pc(S,T) = l4) .
```

The effective condition of this transition is defined by c-assign-v(S,T), which requires that at state S, thread T must be located at l4 (completed loading the shared counter to its local p and its hazard pointer). If that condition is met, the transition can be triggered, and then, in the successor state, the thread moves to l5, and v of T receives the value pointed by p. Otherwise, the state remains unchanged. The values of counter, p, n, and hp of any thread, and values of any pointers never change by this transition regardless the effective condition is met or not.

3.2 Function Reclaim

We additionally introduce some more observation functions to keep track of the detached sets and the thread-local variables inside the function reclaim(). Each of them takes a state and a thread ID as inputs. For instance, detached(S,T) denotes the set detached[T] at state S. The other observation functions capturing the thread-local variables inuse, isfree, m, and i can be understood likewise.

Memory allocation and deallocation. When a thread requests a memory allocation for its pointer n at line l1, an unused memory block will be allocated for the thread. Subsequently, that block of memory must never be used for other allocations. After an object is reclaimed (line f7), the thread releases the associated memory block, and the memory system should allow a new allocation request to use the freed memory block. In order to specify those memory allocation and deallocation operations, we additionally use one more observation function, namely addrInUse, which captures the set of memory addresses that have been allocated but not yet released at a given state S. For each allocation request, a memory will not be allocated unless its address is not in that set. This is part of the effective condition of the transition alloc-n, which specifies the memory allocation for the thread-local pointer n (line l1 of inc()):

```
eq c-alloc-n(S,T,X) = (pc(S,T) = l1 and 0 < T and T <= N and
  not(X \in addrInUse(S))) .
ceq n(alloc-n(S,T,X), T2) = (if T = T2 then X else n(S,T2) fi)
  if c-alloc-n(S,T,X) .
```

When a memory block is released by a thread, it will be removed from the set, allowing a new allocation request to pick it. This is faithfully reflected by the transition free, which attempts to release object m:

```
ceq addrInUse(free(S,T,X)) =
  (if isfree(S,T) then delete(X,addrInUse(S)) else addrInUse(S) fi)
  if c-free(S,T,X) .
eq c-free(S,T,X) = (pc(S,T) = f7 and m(S,T) = X) .
```

4 Reclamation Verification

This section reports the safe memory reclamation verification results of the Shared counter program.

4.1 The First Property

The first property we want to verify is stated as follows: When a thread T attempts to read the value pointed by its local pointer p to assign it to its local v, the value must be a valid integer, i.e., a non-null value. We formalize the property by the following state predicate:

```
-- for all state S and thread T
eq safety1(S,T) = (pc(S,T) = l4)
  implies not(*(S,p(S,T)) = null or p(S,T) = nil) .
```

Precisely, the predicate says that for any state S and any thread T, when T is located at l4, its pointer p and the value pointed by its pointer are non-null. The premise of safety1 may trigger T to proceed with reading the value as it is the effective condition of the transition assign-v as shown in Sect. 3. To verify the property, we prove that safety1 is a valid invariant, i.e., safety1 holds for all state S and all thread T.

Confirming the error when excluding hazard pointer reclamation. Prior to the formal verification of safety1, we have temporarily modified our specification, excluding the safe reclamation based on hazard pointers. Precisely, instead of invoking the function reclaim(p) at line l7, the temporarily revised the specification, letting it directly deallocate p. This will expectedly cause an error as mentioned in Sect. 1, violating what is stated by safety1. Indeed, we pointed out a counterexample of safety1 with respect to the temporarily modified specification, which can be found on the webpage[1]. This also confirms that our specification is executable and eliminates the suspicion that our verification results are achieved simply because the specification is not executed.

Prove by structural induction. To prove that safety1 is invariant, structural induction [17] on the argument of state (s) is used. We need to prove that (1) safety1 holds in all initial states with any thread T, and (2) each transition preserves safety1. The open-close environment is helpful for carrying out such a proof in CafeOBJ. For instance, the proof of (1) is done through the following open-close environment (or open-close fragment):

```
open PROP .
  op t : -> UInt .        -- an arbitrary thread t
  red safety1(init,t) .   -- asks CafeOBJ to reduce it
close
```

where the module PROP contains the complete specification of the program as well as the predicate safety1, and the **red** (standing for reduce) command reduces the given term. If the open-close fragment is executed with CafeOBJ, CafeOBJ will return true, indicating that (1) is proven. A collection of open-close fragments like the above forms the so-called proof scores [17] of safety1.

To complete (2), proof must be given for each transition. With the transition assign-v shown previously, we prove that if safety1 holds in state s, it holds in the successor states assign-v(s,t2) for all t2. The following is the proof attempt of this induction case where the most typical induction hypothesis instance safety1(s,t) is used:

```
open PROP .
  op  s     : -> Sys .        -- an arbitrary state
  ops t t2  : -> UInt .       -- arbitrary threads t and t2
  red safety1(s,t) implies safety1(assign-v(s,t2),t) . -- returns a complex term
close
```

If we execute this open-close fragment with CafeOBJ, the obtained result is neither true nor false, but a composite term. It cannot be reduced because the environment lacks information about t and t2 at state s, for example, whether or not pc(s,t2) is l4. To overcome this situation, typically, case splitting by means of equations is used to split the case into multiple sub-cases and solve each of them. A lemma may be used to solve a case as well. We do not describe this in more detail here, instead, we kindly ask readers to check [17,23].

IPSG. Our verification is assisted by IPSG [22,23], a tool that can automate the proof score writing process. Precisely, providing a CafeOBJ formal specification, state predicates formalizing the properties of interest together with an auxiliary lemma collection, IPSG can generate the proof scores verifying those properties. The tool was implemented by using CafeInMaude [20], a CafeOBJ interpreter

---

[1] https://github.com/duongtd23/Reclamation-CafeOBJ-verification.

implemented in Maude [3,6]. How the tool operates can be briefly summarized as follows. Starting from a collection of open-close fragments, where each of them does not contain any equation and the most typical induction hypothesis instance is used if it is an induction case, IPSG uses Maude metalevel functionalities [3, Chapter 14] to reduce the goal to a term x. If x is neither true nor false, a subterm of x, let's say x', will be chosen by IPSG, and the current case associated with that open-close fragment will be split into two sub-cases: one when x' holds and the other when it does not. The same procedure is applied for each sub-case created until either true or false is returned for the reduction. When false is returned, IPSG tries to discharge the associated case by finding a suitable lemma from the lemma collection provided by human users.

We use IPSG to produce the proof attempt of the property under verification, such as safety1. In such a proof attempt, either true or false is returned for each sub-case. We are supposed to conjecture additional lemmas to discharge the sub-cases in which false is returned. The property and the conjectured lemmas are fed into IPSG, asking it to produce the proof again for the property as well as the proof attempt for those new lemmas. The second proof attempt produced may require us to conjecture some other lemmas. The process is repeated until the final proof contains no sub-case in which false is returned. We believe that using the tool is very helpful for our verification as we only need to concentrate on the creative task - lemma conjecture. The auxiliary lemmas used to complete our verification will be discussed in Sect. 4.3.

4.2 The Second Property

The second property we want to verify is stated as follows: When a thread T is ready to deallocate the memory block occupied by its local m (at line f7 of function reclaim()), the shared counter does not occupy that memory block. This property is formalized by the following predicate:

```
-- for all state S, thread T, and memory address X
eq safety2(S,T,X) = (m(S,T) = X and pc(S,T) = f7 and isfree(S,T))
  implies not(counter(S) = X) .
```

The property is directly inferred from the following lemma - a stronger predicate than safety2:

```
-- for all state S, thread T, and memory address X
eq p6(S,T,X) = (m(S,T) = X) implies not(counter(S) = X) .
```

This lemma states that if m of some thread T is X, then the shared counter must differ from X. To prove this lemma, induction is used with the employment of IPSG.

4.3 Lemmas

Among the lemmas used for the verification, half belong to the function reclaim() and the other half belong to the function inc(). Each of these two lemma groups can be further divided into three sub-groups.

Lemmas for the set of allocated memories. We define and use some lemmas related to the observation function addrInUse:

```
-- for all state S, thread T, and memory address X
eq addruse-lm1(S,X)   = (counter(S) = X) implies X \in addrInUse(S) .
eq rc-addr-lm1(S,T,X) = (m(S,T) = X) implies X \in addrInUse(S) .
eq rc-addr-lm2(S,T,X) = (X \in detached(S,T) or X \in inuse(S,T))
  implies X \in addrInUse(S) .
```

Those lemmas in general state that if a memory block X is pointed by the shared counter or some thread-local variable, then X exists in the set of allocated memories. For example, what is stated by addruse-lm1 is straightforward to understand and conjecture: the memory occupied by the shared counter must be in the set of allocated memories.

Lemmas for the sets detached and inuse. Lemmas related to the sets detached and inuse are classified into this set. They generally state that if a memory block X is pointed by the shared counter or the thread-local variable n or m, then X does not exist in the T's sets detached and inuse; or in some case X does not exist in the two sets of other threads. The following are two lemmas that are classified into this set:

```
-- for all state S, threads T and T2, and memory address X
eq detached-lm1(S,T,X) = (counter(S) = X or n(S,T) = X)
  implies not(X \in detached(S,T) or X \in inuse(S,T)) .
eq rc-detach-lm1(S,T,T2,X) = (m(S,T) = X and not(T = T2))
  implies not(X \in detached(S,T2) or X \in inuse(S,T2)) .
```

Other lemmas. There are some trivial lemmas, such as the following:

```
eq rc-lm5(S,T,X) = not(T <= N) implies not(hp(S,T) = X) .
```

It is trivial due to we only consider N threads with IDs ranging from 1 to N. In other words, our formal specification never touches a thread and its hazard pointer whose ID is greater than N. Despite its triviality, its proof must be made as we are conducting formal verification. Some other lemmas require a well enough understanding of the program and relative creativity to conjecture them, otherwise, we may be stuck to prove them.

5 Verification of Treiber's Stack

We also formally verify the safe memory reclamation of Treiber's stack [24] integrated with the hazard pointer mechanism. This is a concurrent stack implemented by using a linked list, where the head of the list is the top of the stack. Similar to the original Shared counter program, the algorithm is also subjected to the error caused by a thread accessing a memory block that was deallocated by another thread to avoid a memory leak.

```
       struct node {
         int val;
         node *next; };
       node *top = null;
l0   fun pop() {
l1'    node *p, *n; int v;
l1'    while(true) {
l1       hp[tid] = top;
l2       p = hp[tid];
l3       if (p == null) return null;
l4       v = p.val;
l5       n = p.next;
l6       if CAS(&top,p,n) {
l7         reclaim(p); hp[tid] = null;
l7         return v;
l7'      }
l7'    }
l7' }
```
```
l0   fun push(int v) {
l8'    node *nd, *tp;
l8     nd = new(node);
l9     nd.val = v;
l10'   do {
l10      tp = top;
l11      nd.next = tp;
l12    } while(!CAS(&top,tp,nd));
l12' }
```

Fig. 3. Treiber stack [24] with hazard pointers. The function reclaim(int *p) is shown in Fig. 2

Figure 3 depicts the fixed version of the stack using hazard pointers, in which it reuses the function reclaim(int *p) as shown in Fig. 2. The structure node consists of an integer value and a pointer next pointing to the next element of the linked list (or the next element in the stack). The global pointer top, which is the head of the linked list and the top of the stack, is shared by all threads. In the function pop(), without a valid reclamation mechanism, errors may occur when a thread attempts to read the value and the next pointer of its local pointer p at lines l4 and l5. This is similar to what happens with the original Shared counter program as another thread may have already deallocated the top through its local p. If the CAS at line l6 fails, the thread starts the pop procedure again.

We provide modular specifications for Treiber's stack and the Shared counter program. Both specifications use several common parts, such as, the specification for the function reclaim(), the specifications of unsigned integers as well as the execution labels—module LABEL.

We successfully verify the following properties:

1. When a thread T attempts to read the value of its local pointer p to assign it to its local v and n, p must point to a non-null address and the value must be a valid integer. This is the counterpart of the first property we considered in the Shared counter case study.
2. When a thread T is ready to deallocate the memory block occupied by its local m (at line f7 of function reclaim()), no element of the stack occupies that memory block. This also implies that the top of the stack does not occupy the memory block being deallocated.

The verifications of the two programs are also modular in the sense that we could reuse the lemmas for the function `reclaim()` as in the Shared counter program. As Treiber's stack is more complicated than the Shared counter program, the formal verification of the stack is relatively more complicated than that of the Shared counter program.

6 Related Work

Gotsman et al. [9] verified that memory reclamation of the shared counter program integrated with hazard pointers [14], read-copy-update [13], and epoch-based reclamation [7] are safe. The proofs were constructed based on their proposed key concept called *grace period*, i.e., the period of time during which a thread can access certain shared memory blocks without fear that they get deallocated by other threads. However, their proofs were not mechanized by any tools or proof assistants. IPSG assisted our verification by producing the proof scores, which can be executed with CafeOBJ (or CafeInMaude) to confirm the properties under verification. Compared to manual proofs, computer-verified proofs are preferable in the sense that human mistakes lurking in the proofs can be avoided. Since our proof scores can be checked by CafeOBJ or CafeInMaude independently of IPSG, it eliminates the suspicion that bugs might happen in the proof generation by IPSG.

The integration of the hazard pointer mechanism into Treiber's stack was verified in [10,18,21]. Jung et al. [10] used the Iris framework [11] and Coq proof assistant [12] to verify safe memory reclamation of several concurrent data structures integrated with the hazard pointer and read-copy-update mechanisms. Their verification is more advanced than ours in the aspect that they successfully showed that their verification method is applicable to many case studies. Readers are supposed to have enough knowledge of the concurrent separation logic [2] and its set of proof rules to understand their verification. Whereas, our verification relied on induction, a standard and pedagogical proof technique that lets readers easily grasp the basic idea even for those unfamiliar with formal verification.

Parkinson et al. [18] also verified Treiber's stack integrated with hazard pointers in the concurrent separation logic [2]. Their verification approach relied on *history variables* (or ghost variables)—those are additionally introduced to keep track of the status of a memory block (protection and reclamation) and to record events in the past. However, the proof was achieved under the assumption that a no-longer-used memory block previously occupied by the top of the stack could never be picked up by other memory allocation requests. Recall that our specifications allow this scenario, as by the transition `free`, the memory block pointed by pointer m is removed from the set `addrInUse` in the successor state. Furthermore, their verification was not mechanized by any tools or proof assistants.

The integration of the hazard pointer mechanism into Treiber's stack was also verified by Fu et al. [8]. They proposed to extend the rely-guarantee reasoning by introducing past tense temporal operators in the assertion language of the rely-guarantee reasoning so that they no longer need to use history variables to record events in the past as in the work by Parkinson et al. [18]. The verification is done by constructing invariants on the history of execution traces. Similar to the work by Parkinson et al. [18], their verification was not mechanized by any tools or proof assistants.

Also developed on top of CafeInMaude, CafeInMaude Proof Generator & Fixer-upper (CiMPG+F) [19] can produce the proof scripts, which can be then fed into CiMPA—the CafeInMaude Proof Assistant, verifying the invariant properties of interest. CiMPG+F and IPSG complement each other as the two tools produce different kinds of formal proofs accepted by CafeInMaude. However, lemma conjecture tends to be completed more straightforwardly through a proof score than the corresponding proof script because we can explicitly observe the equations characterizing a sub-case in the proof score (more details were discussed in [22,23]). The Maude Interactive Theorem Prover (ITP) [4] allowed users to prove invariant properties w.r.t. Maude specifications. Although ITP offers the auto command to automatically split a goal into multiple subgoals, it is unclear how much automation this command can do. IPSG can automatically conduct case splitting such that true or false is returned for each sub-case generated. We believe that it is very useful as the user conducting verification only needs to concentrate on the lemma conjecture task. Recall again that IPSG produces proof scores, which can be then checked by CafeOBJ/CafeInMaude independently of IPSG.

7 Conclusion

We have formally verified the safe memory reclamation of two concurrent programs, in which the hazard pointer approach is used for memory reclamation. We do not claim that we are the first who verify these two programs. Instead, we verify the safe memory reclamation algorithm in concurrent programs using an alternative technique, that is, by using CafeOBJ and proof scores. The strong point of this verification technique is that the verification follows the same language as the specification task, which makes the entire verification process undergo smoothly, and the verification is based on induction, a standard and pedagogical proof technique that lets readers easily grasp the basic idea even for those unfamiliar with formal verification.

The challenge and also the main weakness of our verification is that many additional lemmas need to be conjectured and used to make the proofs succeed, some of which are trivial like rc-lm5. Nevertheless, through the two verifications, we have learned that in each case study (1) the lemmas for the function reclaim() can be separated and reused; and (2) the lemmas can be divided into three groups, based on which verifier can conjecture lemmas to verify a new case study. We plan to verify the integration of concurrent programs/data structures with some recently advanced mechanisms for safe memory reclamation, such as [15,16], which have not yet been formally verified. With such future work, we aim to complete the verification more efficiently by increasing the modularity and reducing the number of auxiliary lemmas. What has been done in this work is regarded as the initial step towards that goal.

References

1. Astesiano, E., Kreowski, H., Krieg-Brückner, B. (eds.): Algebraic Foundations of Systems Specification. IFIP State-of-the-Art Reports. Springer, Berlin (1999). https://doi.org/10.1007/978-3-642-59851-7
2. Brookes, S.: A semantics for concurrent separation logic. Theor. Comput. Sci. **375**(1–3), 227–270 (2007). https://doi.org/10.1016/J.TCS.2006.12.034
3. Clavel, M., Durán, F., Eker, S., Lincoln, P., Martí-Oliet, N., Meseguer, J., Talcott, C.L. (eds.): All About Maude, Lecture Notes in Computer Science, vol. 4350. Springer, Berlin (2007). https://doi.org/10.1007/978-3-540-71999-1
4. Clavel, M., Palomino, M., Riesco, A.: Introducing the ITP tool: a tutorial. J. Univers. Comput. Sci. **12**(11), 1618–1650 (2006). https://doi.org/10.3217/JUCS-012-11-1618
5. Diaconescu, R., Futatsugi, K.: CafeOBJ Report, AMAST Series in Computing, vol. 6. World Scientific (1998)
6. Durán, F., Eker, S., Escobar, S., Martí-Oliet, N., Meseguer, J., Rubio, R., Talcott, C.L.: Programming and symbolic computation in Maude. J. Log. Algebraic Methods Program. **110** (2020)
7. Fraser, K.: Practical lock-freedom. Ph.D. thesis, University of Cambridge, UK (2004). https://ethos.bl.uk/OrderDetails.do?uin=uk.bl.ethos.599193
8. Fu, M., Li, Y., Feng, X., Shao, Z., Zhang, Y.: Reasoning about optimistic concurrency using a program logic for history. In: Gastin, P., Laroussinie, F. (eds.) CONCUR 2010—Concurrency Theory, 21th International Conference, CONCUR 2010, Paris, France, August 31-September 3, 2010. Proceedings. Lecture Notes in Computer Science, vol. 6269, pp. 388–402. Springer, Berlin (2010). https://doi.org/10.1007/978-3-642-15375-4_27
9. Gotsman, A., Rinetzky, N., Yang, H.: Verifying concurrent memory reclamation algorithms with grace. In: Felleisen, M., Gardner, P. (eds.) Programming Languages and Systems—22nd European Symposium on Programming, ESOP 2013, Held as Part of the European Joint Conferences on Theory and Practice of Software, ETAPS 2013, Rome, Italy, March 16-24, 2013. Proceedings. Lecture Notes in Computer Science, vol. 7792, pp. 249–269. Springer, Berlin (2013). https://doi.org/10.1007/978-3-642-37036-6_15
10. Jung, J., Lee, J., Choi, J., Kim, J., Park, S., Kang, J.: Modular verification of safe memory reclamation in concurrent separation logic. Proc. ACM Program. Lang. **7**(OOPSLA2), 828–856 (2023). https://doi.org/10.1145/3622827
11. Jung, R., Krebbers, R., Jourdan, J., Bizjak, A., Birkedal, L., Dreyer, D.: Iris from the ground up: A modular foundation for higher-order concurrent separation logic. J. Funct. Program. **28**, e20 (2018). https://doi.org/10.1017/S0956796818000151
12. Leroy, X.: Formal verification of a realistic compiler. Commun. ACM **52**(7), 107–115 (2009). https://doi.org/10.1145/1538788.1538814
13. Mckenney, P.E., Walpole, J.: Exploiting deferred destruction: an analysis of read-copy-update techniques in operating system kernels. Ph.D. thesis (2004)
14. Michael, M.M.: Hazard pointers: Safe memory reclamation for lock-free objects. IEEE Trans. Parallel Distributed Syst. **15**(6), 491–504 (2004). https://doi.org/10.1109/TPDS.2004.8
15. Nikolaev, R., Ravindran, B.: Universal wait-free memory reclamation. CoRR **abs/2001.01999** (2020). http://arxiv.org/abs/2001.01999

16. Nikolaev, R., Ravindran, B.: Brief announcement: Crystalline: Fast and memory efficient wait-free reclamation. In: Gilbert, S. (ed.) 35th International Symposium on Distributed Computing, DISC 2021, October 4-8, 2021, Freiburg, Germany (Virtual Conference). LIPIcs, vol. 209, pp. 60:1–60:4. Schloss Dagstuhl - Leibniz-Zentrum für Informatik (2021). https://doi.org/10.4230/LIPICS.DISC.2021.60, https://doi.org/10.4230/LIPIcs.DISC.2021.60
17. Ogata, K., Futatsugi, K.: Proof scores in the OTS/CafeOBJ method. In: FMOODS 2003, France. vol. 2884, pp. 170–184. Springer, Berlin (2003)
18. Parkinson, M.J., Bornat, R., O'Hearn, P.W.: Modular verification of a non-blocking stack. In: Hofmann, M., Felleisen, M. (eds.) Proceedings of the 34th ACM SIGPLAN-SIGACT Symposium on Principles of Programming Languages, POPL 2007, Nice, France, January 17-19, 2007. pp. 297–302. ACM (2007). https://doi.org/10.1145/1190216.1190261
19. Riesco, A., Ogata, K.: Cimpg+f: A proof generator and fixer-upper for cafeobj specifications. In: Theoretical Aspects of Computing - ICTAC 2020—17th International Colloquium, Macau, China, 2020, Proceedings. Lecture Notes in Computer Science, vol. 12545, pp. 64–82. Springer, Berlin (2020). https://doi.org/10.1007/978-3-030-64276-1_4
20. Riesco, A., Ogata, K., Futatsugi, K.: A Maude environment for CafeOBJ. Formal Aspects Comput. **29**(2), 309–334 (2017). https://doi.org/10.1007/S00165-016-0398-7
21. Tofan, B., Schellhorn, G., Reif, W.: Formal verification of a lock-free stack with hazard pointers. In: Cerone, A., Pihlajasaari, P. (eds.) Theoretical Aspects of Computing - ICTAC 2011—8th International Colloquium, Johannesburg, South Africa, 2011. Proceedings. Lecture Notes in Computer Science, vol. 6916, pp. 239–255. Springer, Berlin (2011). https://doi.org/10.1007/978-3-642-23283-1_16
22. Tran, D.D.: Formal verification with algebraic techniques and its application. Ph.D. thesis, Ishikawa 923-1211, Japan (2023). http://hdl.handle.net/10119/18777
23. Tran, D.D., Ogata, K.: Formal verification of TLS 1.2 by automatically generating proof scores. Comput. Secur. **123**, 102909 (2022). https://doi.org/10.1016/j.cose.2022.102909
24. Treiber, R.K.: Systems Programming: Coping with Parallelism. International Business Machines Incorporated, Thomas J. Watson Research Center (2986)

Equivalence, and Property Internalization and Preservation for Equational Programs

José Meseguer[✉]

University of Illinois at Urbana-Champaign, Urbana, IL, USA
meseguer@illinois.edu

Abstract. An equational theory $\mathcal{E} = (\Sigma, E \cup B)$ is an equational program if its equations E, oriented as rewrite rules $\boldsymbol{E}$, are ground convergent modulo axioms B. Its *properties* are the inductive theorems of the initial algebra $\mathbb{T}_{\Sigma/E \cup B}$ defined by $(\Sigma, E \cup B)$. Since programs are structured in module hierarchies, checkable syntactic conditions are given to *preserve* program properties *up and/or down* such hierarchies. Two equational programs $\mathcal{E} = (\Sigma, E \cup B)$ and $\mathcal{E}' = (\Sigma, E' \cup B')$ are *equivalent* iff they define the same computable functions on the same algebraic data types. Succinct conditions to verify $\mathcal{E}$ and $\mathcal{E}'$ equivalent are given. A useful *internalization* method to extend an equational program $\mathcal{E}$ into an equivalent one by adding new rewrite rules or structural axioms that are inductive theorems of $\mathcal{E}$ is also given. This method can make proofs of program properties simpler and shorter, and offers a new way to prove equational theories ground convergent.

1 Introduction

Equational programming is a declarative, functional programming paradigm where programs are equational theories $\mathcal{E} = (\Sigma, E \cup B)$ whose equations E, oriented as rewrite rules $\boldsymbol{E}$, are ground confluent modulo axioms B like associativity and/or commutativity and/or unit element, thus ensuring *determinism*, i.e., a unique result if the program terminates. Reasoning about equational program properties is much easier than doing so for imperative programs, because such properties are the first-order formulas valid in the initial algebra $\mathbb{T}_{\Sigma/E \cup B}$ defined by $(\Sigma, E \cup B)$ or, assuming the rules $\boldsymbol{E}$ are terminating modulo B (which I will do throughout), its isomorphic *canonical term algebra* $\mathbb{C}_{\Sigma/E,B}$. The expressiveness of equational programs is greatly enhanced by writing them in an equational logic supporting types and subtypes. In this paper I assume that they are order-sorted equational theories [10], the logic chosen by the OBJ [11], CafeOBJ [8] and Maude [4] languages, which can be further extended to membership equational logic (MEL) [17]. The expressiveness of rewriting modulo axioms B can be further increased by: (i) supporting conditional equations; (ii) supporting parametric polymorphism through parameterized theories; (iii) exploiting the fact that MEL and its order-sorted sublogic are reflective [1]; and (iv) supporting symbolic computation with logical variables. Maude supports all features (i)–(iv).

I present new concepts, results and techniques to make the verification of order-sorted equational programs modulo structural axioms more modular, succinct and reusable. After some preliminaries in Sect. 2, this is done as follows:

1. In Sect. 3 I study how equational program properties can be preserved *up and/or down* program module hierarchies, and provide checkable, syntactic conditions ensuring the correctness of such property preservations.
2. In Sect. 4 I propose a natural notion of equational program *equivalence*. Equational logic is a very high-level language for defining recursive functions between user-definable algebraic data types. An equational program $\mathcal{E} = (\Sigma, E)$ is, simultaneously, an *algorithm* for defining and computing with recursive functions for the function symbols in Σ, and a *definition* of the computable data types of their domains and ranges. The most obvious definition of program equivalence between $\mathcal{E}$ and $\mathcal{E}'$ is that they define the *same* recursive functions on the *same* algebraic data types. I give succinct conditions for verifying equational program equivalence by inductive theorem proving, and to preserve program equivalence under reducts.
3. In Sect. 5 I propose the notion of *internalization* of equations G whose conjunction is an inductive theorem of an equational program $\mathcal{E}$ as new rewrite rules or structural axioms that, when added to ("internalized in") $\mathcal{E}$, give rise to a new equational program equivalent to $\mathcal{E}$; and I give succinct conditions enabling such internalizations. This gives rise to an *internalize and conquer* methodology to build long chains of program equivalences, and to make trivial the proofs of equivalence for other programs that can be "interpolated" within such chains. I show how this method, which is supported by Maude's **NuITP** inductive theorem prover [3], can make inductive proofs of equational program properties simpler and shorter, and provides a new method for proving that an equational theory $\mathcal{E}$ is ground confluent.
4. In Sect. 6 I discuss related work and present some conclusions. Proofs of all theorems and lemmas are included in Appendix A. All **NuITP** proof scripts can be found in Appendix B.

To the best of my knowledge, except for the folklore Theorems 1 and 3, all other theorems and lemmas are new.

2 Preliminaries

I assume familiarity with the notions of an order-sorted signature Σ on a poset of sorts $(S, \leq)$, an order-sorted Σ algebra $\mathbb{A}$, and the term Σ-algebras $\mathbb{T}_\Sigma$ and $\mathbb{T}_\Sigma(X)$ for X an S-sorted set of variables. I also assume familiarity with the notions of: (i) Σ-homomorphism $h : \mathbb{A} \to \mathbb{B}$ between Σ-algebras $\mathbb{A}$ and $\mathbb{B}$, so that Σ-algebras and Σ-homomorphisms form a category $\mathbf{OSAlg}_\Sigma$; (ii) order-sorted (i.e., sort-preserving) substitution θ, its domain $dom(\theta)$ and range $ran(\theta)$, and its application $t\theta$ to a term t; (iii) *preregular* order-sorted signature Σ, i.e., a signature such that each term t has a least sort, denoted $ls(t)$; (iv) the set $\widehat{S} = S/(\geq \cup \leq)^+$ of *connected components* of a poset $(S, \leq)$ viewed as a DAG;

(v) operators $f : s_1 \ldots s_n \to s$ and $f : s'_1 \ldots s'_n \to s'$ in Σ are called *subsort-overloaded* iff $[s_1] = [s'_1], \ldots, [s_n] = [s'_n], [s] = [s']$, where $[s''] \in \widehat{S}$ denotes the connected component of $s \in S$; (vi) for $\mathbb{A}$ a Σ-algebra, the sets A_s of its elements of sort $s \in S$ and for each $[s] \in \widehat{S}$ the set $A_{[s]} = \bigcup_{s' \in [s]} A_{s'}$; (vii) the order-sorted equational deduction relation $E \vdash u = v$ and its associated E-equality relation $=_E$; (viii) the *initial algebra* $\mathbb{T}_{\Sigma/E}$ associated to an equational theory (Σ, E), where, for each $s \in S$, $T_{\Sigma/E,s} = \{[t] \in T_{\Sigma,[s]}/=_E \mid [t] \cap T_{\Sigma,s} \neq \emptyset\}$, and for each $f : s_1 \ldots s_n \to s \in \Sigma$, $f_{\mathbb{T}_{\Sigma/E}}([t_1], \ldots, [t_n]) = [f(t'_1, \ldots, t'_n)]$, where $t'_i \in [t_i] \cap T_{\Sigma,s_i}$, $1 \leq i \leq n$; (ix) E-*unifiers* of an equation $u = v$, i.e., substitutions θ such that $u\theta =_E v\theta$, and complete set $Unif_E(u = v)$ of E-unifiers; and (x) the satisfaction relation $\mathbb{A} \models \varphi$ of a first-order Σ-formula φ by a Σ-algebra $\mathbb{A}$. I furthermore assume that all signatures Σ have *non-empty sorts*, i.e., $T_{\Sigma,s} \neq \emptyset$ for each $s \in S$. $[A \to B]$ denotes the *S-sorted functions* from A to B. These notions are explained in [10,17,18]. The material below is adapted from [14,18].

Convergent Theories and Sufficient Completeness. Given an order-sorted equational theory $\mathcal{E} = (\Sigma, E \cup B)$, where B is a collection of associativity (A) and/or commutativity (C) and/or unit element (U) axioms and Σ is B-preregular,[1] we can associate to it a corresponding *rewrite theory* [16] $\vec{\mathcal{E}} = (\Sigma, B, \vec{E})$ by orienting the equations E as left-to-right rewrite rules. That is, each $(u = v) \in E$ is transformed into a rewrite rule $u \to v$. For simplicity I recall here the case of unconditional equations and refer to [14] for full details on the general definition of convergent *conditional* theory $\mathcal{E}$. The rewrite theory $\vec{\mathcal{E}}$ reduces the complex bidirectional reasoning with equations to the much simpler unidirectional reasoning with rules under suitable assumptions. I assume familiarity with the notion of subterm $t|_p$ of t at a term position p and of term replacement $t[w]_p$ of $t|_p$ by w at position p (see, e.g., [2]). The rewrite relation $t \to_{\vec{E},B} t'$ holds iff there is a subterm $t|_p$ of t, a rule $(u \to v) \in \vec{E}$ and a substitution θ such that $u\theta =_B t|_p$ and $t' = t[v\theta]_p$. The reflexive-transitive closure of $\to_{\vec{E},B}$ is denote by $\to^*_{\vec{E},B}$. For $\mathcal{E}$ unconditional, the convergence requirements are as follows (see [14] for $\mathcal{E}$ conditional): (i) $vars(v) \subseteq vars(u)$; (ii) *sort-decreasingness*: for each substitution θ, $ls(u\theta) \geq ls(v\theta)$; (iii) *strict B-coherence*: if $t_1 \to_{\vec{E},B} t'_1$ and $t_1 =_B t_2$ then there exists $t_2 \to_{\vec{E},B} t'_2$ with $t'_1 =_B t'_2$; (iv) *confluence* (resp. *ground confluence*) *modulo B*: for each term t (resp. ground term t) if $t \to^*_{\vec{E},B} v_1$ and $t \to^*_{\vec{E},B} v_2$, then there exist rewrite sequences $v_1 \to^*_{\vec{E},B} w_1$ and $v_2 \to^*_{\vec{E},B} w_2$ such that $w_1 =_B w_2$; (v) *termination*: the relation $\to_{\vec{E},B}$ is well-founded (for $\vec{E}$ conditional, we require *operational termination* [14]). If $\mathcal{E}$ satisfies conditions (i)–(v) (resp. the same, but (iv) weakened to ground confluence modulo B), then it is called *convergent* (resp. *ground convergent*). The key point is that then, given a term (resp. ground term) t, all terminating rewrite sequences $t \to^*_{\vec{E},B} w$ end in a term w, denoted $t!_\mathcal{E}$, that is unique up to B-equality, and its called t's *canonical form*. Ground convergence implies three

---

[1] A signature Σ is *B-preregular* if it is preregular and for each $[t] \in T_{\Sigma/B}$ there is an effectively determined least sort s such that $[t] \in T_{\Sigma/B,s}$ (if B has only A and/or C axioms, this is the least sort of any $t' \in [t]$). This property is checked by Maude.

major results: (1) for any ground terms t, t' we have $t =_{E \cup B} t'$ iff $t!_\mathcal{E} =_B t'!_\mathcal{E}$, (2) the B-equivalence classes of canonical forms are the elements of the *canonical term algebra* $\mathbb{C}_{\Sigma/E,B}$, where for each $f : s_1 \ldots s_n \to s$ in Σ and B-equivalence classes of canonical terms $[t_1], \ldots, [t_n]$ with $ls(t_i) \leq s_i$ the operation $f_{\mathbb{C}_{\Sigma/E,B}}$ is defined by the identity: $f_{\mathbb{C}_{\Sigma/E,B}}([t_1] \ldots [t_n]) = [f(t_1 \ldots t_n)!_\mathcal{E}]$, and (3) we have an isomorphism $\mathbb{T}_{\Sigma/E \cup B} \cong \mathbb{C}_{\Sigma/E,B}$.

A ground convergent rewrite theory $\mathcal{E} = (\Sigma, B, E)$ is called *sufficiently complete* with respect to a subsignature Ω, whose operators are called *constructors*, iff for each ground Σ-term t, $t!_\mathcal{E} \in T_\Omega$. Define Ω^+ as Ω's subsort-overloaded closure. That is, for any $c : s_1 \ldots s_n \to s \in \Sigma$, $c : s_1 \ldots s_n \to s \in \Omega^+$ iff

$$\exists c : s'_1 \ldots s'_n \to s' \in \Omega \text{ s.t. } [s_1] = [s'_1], \ldots, [s_n] = [s'_n], [s] = [s'].$$

Note that Ω^+ splits Σ into a disjoint union $\Sigma = \Delta \uplus \Omega^+$ with $\Delta = \Sigma \setminus \Omega^+$, with no subsort overloading possible between function symbols in Δ and in Ω^+. For $\mathcal{E} = (\Sigma, B, E)$ sufficiently complete w.r.t. Ω, a ground convergent rewrite subtheory $(\Omega^+, B_{\Omega^+}, E_{\Omega^+}) \subseteq (\Sigma, B, E)$ is called a *constructor subspecification* iff[2] $\mathbb{C}_{\Sigma/E,B}|_{\Omega^+} \cong \mathbb{C}_{\Omega^+/E_{\Omega^+},B_{\Omega^+}}$. If E_{Ω^+} is such that each $u \in T_\Omega$ is in $E_{\Omega^+}, B_{\Omega^+}$-normal form, Ω is called a signature of *free constructors modulo axioms* B_Ω.

Example 1. The Maude functional module:

```
fmod NAT+AC is
  sorts Natural NzNatural .  subsort NzNatural < Natural .
  op 0 : -> Natural [ctor] .
  op 1 : -> NzNatural [ctor] .
  op _+_ : Natural Natural -> Natural [assoc comm] .
  op _+_ : NzNatural NzNatural -> NzNatural [ctor assoc comm] .
  eq N:Natural + 0 = N:Natural .
endfm
```

with signature Ω^+ and subsignature $\Omega \subset \Omega^+$ the symbols with the `ctor` attribute, has Ω as signature of free constructors modulo AC. The free constructor modulo AC Ω-terms are: 0, 1, and $1+ \overset{k}{\ldots} +1$, $k > 1$.

3 Modular Preservation of Program Properties

Not every order-sorted equational theory $\mathcal{E} = (\Sigma, E \cup B)$ is suitable as an *equational program* executable by term rewriting. Call $\mathcal{E} = (\Sigma, E \cup B)$ *admissible* (as a program) iff $\mathcal{E} = (\Sigma, B, E)$ is ground convergent. It is useful to make explicit the constructor subtheory $(\Omega^+, B_{\Omega^+}, E_{\Omega^+}) \subseteq (\Sigma, B, E)$ on which $\mathcal{E}$ is *sufficiently complete* w.r.t. $\Omega \subseteq \Omega^+$. Talk about the *formal properties* of an equational program $\mathcal{E} = (\Sigma, E \cup B)$ is meaningless if no formal semantics has been given to it. Since $\mathcal{E}$ is ground convergent, I propose as its semantics the Σ-algebra $\mathbb{C}_{\Sigma/E,B}$.

---

[2] For a Σ-algebra $\mathbb{A}$ and a subsignature Σ' with same poset of sorts, the *reduct* $\mathbb{A}|_{\Sigma'}$ is the Σ'-algebra with same sorts as $\mathbb{A}$ and same operations $f_\mathbb{A}$ as $\mathbb{A}$ for each $f \in \Sigma'$.

This semantics is very precise, considerably more so than a vanilla-flavored "initial algebra semantics," because $\mathbb{C}_{\Sigma/E,B}$ specifies the *computable data types*, and the *computable functions* defined by the *functional program* $\mathcal{E}$. Technical details become simpler if (as I shall do henceforth) we assume that in $B = B_\Delta \uplus B_{\Omega^+}$ the unit element axioms of B only occur in B_{Ω^+}, so that all axioms in B_Δ are associative and/or commutative. Likewise, the rules E split as a disjoint union $E = E_\Delta \uplus E_{\Omega^+}$, where I shall always assume that the rules in E_Δ have the form $f(u_1, \ldots, u_n) \to v$ *if* φ for some $f \in \Delta$. Our assumption on B_Δ implies that the Ω^+-algebra isomorphism $\mathbb{C}_{\Sigma/E,B}|_{\Omega^+} \cong \mathbb{C}_{\Omega^+/E_{\Omega^+},B_{\Omega^+}}$ becomes the Ω^+-algebra equality $\mathbb{C}_{\Sigma/E,B}|_{\Omega^+} = \mathbb{C}_{\Omega^+/E_{\Omega^+},B_{\Omega^+}}$. Since $\mathbb{C}_{\Sigma/E,B}|_{\Omega^+} = \mathbb{C}_{\Omega^+/E_{\Omega^+},B_{\Omega^+}}$, the data types of the functional program $\mathcal{E}$ are obviously the *computable data types* $\{C_{\Omega^+/E_{\Omega^+},B_{\Omega^+},s}\}_{s\in S}$ whose elements are B_{Ω^+}-equivalence classes of constructor terms in $E_{\Omega^+}, B_{\Omega^+}$-canonical form. For each $f : s_1 \ldots s_n \to s$ in Σ, the *function* defined by $\mathcal{E}$ for f is obviously the *computable function* $f_{\mathbb{C}_{\Sigma/E,B}} : C_{\Omega^+/E_{\Omega^+},B_{\Omega^+},s_1} \times \ldots \times C_{\Omega^+/E_{\Omega^+},B_{\Omega^+},s_n} \to C_{\Omega^+/E_{\Omega^+},B_{\Omega^+},s}$.

The above equational program semantics provides an immediate answer to the question: What are the formal properties of the functional equational program $\mathcal{E}$? Obviously, those of the family of *computable functions* defined by $\mathcal{E}$, which can be naturally expressed as first-order Σ-formulas φ. Therefore, program $\mathcal{E}$ satisfies property φ iff, by definition, $\mathbb{C}_{\Sigma/E,B} \models \varphi$. Since $\mathbb{T}_{\Sigma/E \cup B} \cong \mathbb{C}_{\Sigma/E,B}$, φ is the property of an initial algebra, and therefore an *inductive* property [19]. Therefore, equational program verification is just inductive theorem proving. Furthermore, I will show in Sects. 4 and 5 that inductive theorem proving allows verifying, not just properties of an equational program $\mathcal{E}$, but also *equivalence* between two equational programs $\mathcal{E}$ and $\mathcal{E}'$.

In this section I focus on *modular* preservation of equational program properties. Programs should be structured in *module hierarchies*, where each link in such a hierarchy is an equational *theory inclusion* $\mathcal{E} \subseteq \mathcal{E}'$ and, more precisely, a rewrite theory inclusion $\mathcal{E} = (\Sigma, B, E) \subseteq (\Sigma', B', E') = \mathcal{E}'$. Any such inclusion defines a *unique* Σ-homomorphism $h : \mathbb{C}_{\Sigma/E,B} \to \mathbb{C}_{\Sigma'/E',B'}|_\Sigma$. This is because $B \subseteq B'$ and $E \subseteq E'$, and therefore $\mathbb{C}_{\Sigma'/E',B'}|_\Sigma \models E \cup B$, so that h is the unique Σ-homomorphism from the initial $\mathcal{E}$-algebra $\mathbb{C}_{\Sigma/E,B}$. Modular preservation of equational program properties is thus a special case of the well-known preservation of first-order formulas under various kinds of homomorphisms. I will focus on three kinds of module inclusions, specified in Maude with the respective keywords **extending**, **generated-by** and **protecting**. In the argot of algebraic specifications they characterize module inclusions where the supermodule creates, respectively, "no confusion," "no junk," and "no junk and no confusion" on the submodule. Their meanings for a module inclusion $\mathcal{E} \subseteq \mathcal{E}'$ is that the unique Σ-homomorphism $h : \mathbb{C}_{\Sigma/E,B} \to \mathbb{C}_{\Sigma'/E',B'}|_\Sigma$ is, respectively, *injective*, *surjective*, or *bijective*. By "injective," resp. "surjective," resp. "bijective," I mean that for each sort s in Σ the function h_s is so. Regarding program properties, I shall restrict myself to quantifier-free (QF) formulas φ or, equivalently, to their universal closures $\forall \varphi$. The respective formula preservation results are:

Theorem 1. (Up and Down). *Let $\mathcal{E} \subseteq \mathcal{E}'$ be an inclusion of admissible equational programs where the signature Σ of $\mathcal{E}$ is such that each connected component $[s]$ of its poset of sorts $(S, <)$ has a top element. Then,*

1. **Down**: *If the inclusion is* **extending**, *then for any QF Σ-formula φ,*
   $\mathbb{C}_{\Sigma'/E',B'} \models \varphi \Rightarrow \mathbb{C}_{\Sigma/E,B} \models \varphi$.
2. **Up**: *If the inclusion is* **generated-by**, *then for any positive[3] QF formula φ,*
   $\mathbb{C}_{\Sigma/E,B} \models \varphi \Rightarrow \mathbb{C}_{\Sigma'/E',B'} \models \varphi$.
3. **Up and Down**: *If the inclusion is* **protecting**, *then for any QF Σ-formula φ, $\mathbb{C}_{\Sigma'/E',B'} \models \varphi \Leftrightarrow \mathbb{C}_{\Sigma/E,B} \models \varphi$.*

The above theorem follows from well-known results in Model Theory (see, e.g., Sect. 2.4 in [13]) under the above assumption on Σ. **Down** follows from the Los-Tarski Theorem because $\mathbb{C}_{\Sigma/E,B}$ is isomorphic to a Σ-subalgebra of $\mathbb{C}_{\Sigma'/E',B'}|_{\Sigma}$. **Up** follows from Lyndon's Positivity Theorem. **Up and Down** follows from bijective Σ-homomorphisms being isomorphims under the above assumption on Σ, which preserve all formulas. Theorem 1 is useful in equational program verification because it allows *reuse* of a property φ proved for one module as a property of a sub- or super-module, provided appropriate requirements on φ and on the module inclusion hold. But, of course, property reuse hinges on verifying that a submodule inclusion $\mathcal{E} \subseteq \mathcal{E}'$ is actually **extending**, resp. **generated-by**, resp. **protecting**. Before addressing this issue, let me first recall the following results from [5].

Given a ground convergent theory $\mathcal{E} = (\Sigma, B, E)$ with B any combination of associativity, commutativity and unit axioms, B decomposes as a disjoint union $B = B_{A \vee C} \uplus U$, with $B_{A \vee C}$ the associativity and/or commutativity axioms and U the unit axioms. Furthermore, (i) any unit axiom $f(x, e) = x$ (resp. $f(e, x) = x$) can be oriented as a rule $f(x, e) \to x$, (resp. $f(e, x) \to x$), yielding a set U of rules such that $(\Sigma, B_{A \vee C}, U)$ is convergent and enjoys the finite variant property (FVP) [7]; (ii) therefore, for any terms u, v by the Church-Rosser Theorem, $u =_B v$ iff $u!_{U,B_{A\vee C}} =_{B_{A\vee C}} v!_{U,B_{A\vee C}}$; (iii) $\mathcal{E}$ is semantically equivalent to the ground convergent theory $\mathcal{E}_U = (\Sigma, B_{A\vee C}, E_U \uplus U)$, where E_U can be automatically generated from E by computing for each $l \to r$ in E the finite set of variants of its lefthand side l (see [5]); (iv) the $\to_{E,B}$, $\to_{U,B_{A\vee C}}$ and $\to_{E_U,B_{A\vee C}}$ relations are related as follows: (iv).1 $u \to_{E,B} v \Rightarrow \exists v'$ s.t. $u!_{U,B_{A\vee C}} \to_{E_U,B_{A\vee C}} v' \wedge v =_B v'$, and (iv).2. $u!_{U,B_{A\vee C}} \to_{E_U,B_{A\vee C}} v' \Rightarrow \exists v$ s.t. $u \to_{E,B} v \wedge v =_B v'$. The following theorem gives checkable conditions ensuring module inclusions to be **extending**, resp. **generated-by**, resp. **protecting**. Its proof is given in Appendix A.

Theorem 2. *Let $\mathcal{E} = (\Sigma, B, E) \subseteq (\Sigma', B', E') = \mathcal{E}'$ be an inclusion of admissible equational theories sufficiently complete w.r.t. Ω, resp. Ω', satisfies the assumptions in Theorem 1, and is such that: (i) any subsort-overloaded operators $f : s_1 \ldots s_n \to s$ in Σ and $f : s'_1 \ldots s'_n \to s'$ in Σ' have the same*

---

[3] By definition, φ is *positive* iff it is Boolean-equivalent to a QF whose only Boolean connectives are $\vee$ and $\wedge$.

axioms in B and in B', (ii) the corresponding inclusion of constructor subtheories $\mathcal{E}_{\Omega^+} = (\Omega^+, B_{\Omega^+}, E_{\Omega^+}) \subseteq \mathcal{E}'_{\Omega'^+} = (\Omega'^+, B'_{\Omega'^+}, E'_{\Omega'^+})$ is such that the rules in $E'_{\Omega'^+}$ are unconditional and $\Delta \subseteq \Delta'$, and (iii) as noted above, $B'_{\Omega'^+}$ and B_{Ω^+} both decompose as disjoint unions $B'_{\Omega'^+} = B'_{A\vee C,\Omega'^+} \uplus U_{\Omega'^+}$ and $B_{\Omega^+} = B_{A\vee C,\Omega^+} \uplus U_{\Omega^+}$. Then,

1. The inclusion is **extending** if for any rule $l' \to r'$ in $E'_{\Omega'^+ \cup U_{\Omega'^+}}$ and for any sort specialization[4] ρ such that $l\rho$ is an Ω^+-term there is a rule $l \to r$ in $E_{\Omega^+ \cup U_{\Omega^+}}$ and a sort specialization τ such that the rules $l'\rho \to r'\rho$ and $l\tau \to r\tau$ are identical.
2. The inclusion is **generated-by** if (a) $\forall s' \in S'$, $\forall s \in S, (s' <' s) \Rightarrow (s' \in S \wedge s' < s)$, and (b) for any $c : s'_1 \ldots s'_n \to s'$ in Ω'^+ such that $s' \in S$ we must have $s'_1, \ldots, s'_n \in S$ and $c : s'_1 \ldots s'_n \to s'$ in Ω^+.
3. The inclusion is **protecting** if it satisfies the conditions in both (1) and (2).

To the best of my knowledge, Theorem 2 is new. It has the advantage of making the conditions for checking module inclusions to be **extending**, resp. **generated-by**, resp. **protecting**, *purely syntactic* and therefore automatable. This facilitates the reusability of equational program properties afforded by Theorem 1. It is also very widely applicable, since condition (i) is enforced by OBJ3, CafeOBJ and Maude on all module inclusions, and condition (ii) on constructor equations being unconditional is very often satisfied in practice.

Example 2. Consider the following inclusion of Maude equational programs in which ACU addition is extended to AC multiplication:

```
fmod NAT+ACU is
  sorts Natural NzNatural .    subsort NzNatural < Natural .
  op 0 : -> Natural [ctor] .
  op 1 : -> NzNatural [ctor] .
  op _+_ : Natural Natural -> Natural [assoc comm id: 0] .
  op _+_ : NzNatural NzNatural -> NzNatural [ctor assoc comm id: 0] .
endfm

fmod NAT+ACU*AC is protecting NAT+ACU .
  op _*_ : Natural Natural -> Natural [assoc comm] .
  vars N M K : Natural .    vars N' M' : NzNatural .
  eq N * 0 = 0 .
  eq N * 1 = N .
  eq K * (N' + M') = (K * N') + (K * M') .
endfm
```

Both NAT+ACU and NAT+ACU*AC are convergent, and the subsignature Ω of NAT+ACU specified by the ctor keyword makes both modules sufficiently complete w.r.t. Ω. We can use Theorem 2 to check that the stated **protecting** inclusion is indeed **protecting**. Condition (1) holds trivially, since NAT+ACU*AC contains

---
[4] A *sort specialization* is a bijective substitution $\rho = \{x_1 \mapsto x'_1, \ldots x_n \mapsto x'_n\}$ such that if x_i has sort s_i, then x'_i has sort s'_i with $s_i \geq s'_i$, $1 \leq i \leq n$.

no new constructors (actually, NAT+ACU is its constructor subspecification). Condition (2) also holds trivially, since both modules share the same constructor subspecification.

4 Program Equivalence

Before discussing theory and program equivalences I recall a folklore theorem on initial algebras. For the sake of self-containedness its proof is given in Appendix A. By definition, the *inductive consequence relation* $(\Sigma, E) \models_{ind} \Gamma$ holds between an equational theory (Σ, E) and a set Γ of first-order Σ-formulas iff $\mathbb{T}_{\Sigma/E} \models \Gamma$, i.e., if all $\varphi \in \Gamma$ are *inductive theorems* of the initial algebra $\mathbb{T}_{\Sigma/E}$.

Theorem 3. *Let (Σ, E) be an order-sorted equational theory and G a set of Σ-equations. Then, $(\Sigma, E) \models_{ind} G$ iff $\mathbb{T}_{\Sigma/E} = \mathbb{T}_{\Sigma/E \cup G}$.*

Inductive Equivalence. Call two equational theories (Σ, E) and (Σ, E') *inductively equivalent*, denoted $(\Sigma, E) \equiv_{ind} (\Sigma, E')$, iff $\mathbb{T}_{\Sigma/E} = \mathbb{T}_{\Sigma/E'}$, and therefore both have the same inductive theorems. This defines an equivalence relation between equational theories. Note that, by Theorem 3 and transitivity and symmetry of $\equiv_{ind}$, $(\Sigma, E) \equiv_{ind} (\Sigma, E')$ iff $(\Sigma, E) \models_{ind} E'$ and $(\Sigma, E') \models_{ind} E$.

Program Equivalence. When are two equational programs, i.e., two admissible theories $\mathcal{E} = (\Sigma, E \cup B)$ and $\mathcal{E}' = (\Sigma, E' \cup B')$, both sufficiently complete w.r.t. constructors Ω, *equivalent*? The most obvious answer is: when they define the *same functions* on the *same data types*, which exactly means that $\mathbb{C}_{\Sigma/E,B} = \mathbb{C}_{\Sigma/E',B'}$. This defines an equivalence relation on equational programs which I call *program (semantic) equivalence*, denoted $\mathcal{E} \equiv_{sem} \mathcal{E}'$. Obviously, $\mathcal{E} \equiv_{sem} \mathcal{E}'$ implies $\mathcal{E} \equiv_{ind} \mathcal{E}'$, but the converse implication does not hold in general. Note that it follows from the very definition of $\mathcal{E} \equiv_{sem} \mathcal{E}'$ (and also from the weaker property $\mathcal{E} \equiv_{ind} \mathcal{E}'$) that programs $\mathcal{E}$ and $\mathcal{E}'$ satisfy exactly the same properties φ. This simple observation provides what might be called an *horizontal property preservation principle*,[5] since we can automatically *transfer* any inductive theorems proved for $\mathcal{E}$ to $\mathcal{E}'$ and conversely. The proof of Lemma 1 is in Appendix A.

Lemma 1. *Let $\mathcal{E} = (\Delta \uplus \Omega^+, E \cup B)$ and $\mathcal{E}' = (\Delta \uplus \Omega^+, E' \cup B')$ be admissible, sufficiently complete w.r.t. constructors Ω and semantically equivalent; and let $\Delta_0 \subseteq \Delta$ be a subsignature closed in Δ under subsort overloading such that for all rules $f(u_1, \ldots, u_n) \to v$ if φ in the subset $\boldsymbol{E}_{\Delta_0}$ of $\boldsymbol{E}_\Delta$ (resp. $\boldsymbol{E}'_{\Delta_0}$ of $\boldsymbol{E}'_\Delta$) such that $f \in \Delta_0$, $f(u_1, \ldots, u_n)$, v and the terms in φ are $\Delta_0 \uplus \Omega^+$-terms. Let B_{Δ_0} (resp. B'_{Δ_0}) denote the subset of B_Δ (resp. B'_Δ) involving a binary $f \in \Delta_0$. Then, $(\Delta_0 \uplus \Omega^+, E_{\Delta_0} \cup E_{\Omega_+} \cup B_{\Delta_0} \cup B_{\Omega^+})$ and $(\Delta_0 \uplus \Omega^+, E'_{\Delta_0} \cup E'_{\Omega_+} \cup B'_{\Delta_0} \cup B'_{\Omega^+})$ are both admissible, sufficiently complete w.r.t. Ω and semantically equivalent.*

---

[5] Horizontal, since no submodule inclusions need exist between $\mathcal{E}$ and $\mathcal{E}'$.

Since Alan Turing, the (undecidable) problem of verifying program equivalence has been a central issue in program verification. How can we verify that two equational programs $\mathcal{E}$ and $\mathcal{E}'$ are semantically equivalent? For the sake of a simpler exposition I answer this question in the theorem below, whose proof is given in Appendix A, when the equations in $\mathcal{E}$ and $\mathcal{E}'$ are *unconditional*. The general case allowing conditional equations will be treated elsewhere.

Theorem 4. (Program Equivalence). *For admissible unconditional equational programs* $\mathcal{E} = (\Sigma, E \cup B)$ *and* $\mathcal{E}' = (\Sigma, E' \cup B')$ *sufficiently complete w.r.t.* Ω, *resp.* Ω', *and with respective constructor subspecifications* $(\Omega^+, E_{\Omega^+} \cup B_{\Omega^+})$ *and* $(\Omega^+, E'_{\Omega^+} \cup B'_{\Omega^+})$, $\mathcal{E} \equiv_{sem} \mathcal{E}'$ *iff:*

1. $\mathbb{C}_{\Omega^+/E_{\Omega^+},B_{\Omega^+}} = \mathbb{C}_{\Omega^+/E'_{\Omega^+},B'_{\Omega^+}}$, *and*
2. $(\Sigma, E \cup B) \models_{ind} (E'_\Delta \setminus E_\Delta) \cup (B'_\Delta \setminus B_\Delta)$.

Note the economy of proof afforded by the remarkably asymmetric condition (2) for checking the symmetric relation $\equiv_{sem}$. The user can choose either $\mathcal{E}$ or $\mathcal{E}'$ as the ground for the inductive proofs in (2), whichever choice makes proofs easier. Of course, Theorem 4 still leaves open the question of how to verify condition (1) that $\mathbb{C}_{\Omega^+/E_{\Omega^+},B_{\Omega^+}} = \mathbb{C}_{\Omega^+/E'_{\Omega^+},B'_{\Omega^+}}$. This question is answered in Theorem 7. But in the very common case when $(\Omega^+, E_{\Omega^+} \cup B_{\Omega^+}) = (\Omega^+, E'_{\Omega^+} \cup B'_{\Omega^+})$ (1) holds trivially and only (2) needs to be checked.

Example 3. This example continues Example 2 by considering a different definition of natural number addition and multiplication in which addition is defined as *string* (associative) concatenation (indeed, as counting with one's fingers):

```
fmod NAT+AU is
  sorts Natural NzNatural .   subsort NzNatural < Natural .
  op 0 : -> Natural [ctor] .
  op 1 : -> NzNatural [ctor] .
  op _+_ : Natural Natural -> Natural [assoc id: 0] .
  op _+_ : NzNatural NzNatural -> NzNatural [ctor assoc id: 0] .
endfm

fmod NAT+AU* is protecting NAT+AU .
  op _*_ : Natural Natural -> Natural .
  vars N M K : Natural .   var N' : NzNatural .
  eq N * 0 = 0 .
  eq N * 1 = N .
  eq K * (N' + 1) = K + (K * N') .
endfm
```

By Theorem 4, NAT+ACU*AC $\equiv_{sem}$ NAT+AU* holds iff we prove: (1) NAT+ACU $\equiv_{sem}$ NAT+AU, and (2) NAT+ACU*AC $\models K * (N' + 1) = K + (K * N')$. (1) is proved in Example 4. The NuITP script proving (2) can be found in Appendix B.

5 Internalizing Program Properties

Program properties that are equations which can be added to the program as terminating rules or as axioms extend such a program to a semantically equivalent one by a process that I call *internalization*. Theorem 5's proof is in Appendix A.

Theorem 5. (Rule Internalization). *Let $(\Sigma, E \cup B)$ be an admissible equational program sufficiently complete w.r.t. Ω and with constructor subspecification $(\Omega^+, E_{\Omega^+} \cup B_{\Omega^+})$; and let G be a finite set of Σ-equations such that: (i) $(\Sigma, E \cup B) \models_{ind} G$, (ii) the equations G can be oriented as sort-decreasing rules $\boldsymbol{G}$ of either the form $f(u_1, \ldots, u_n) \to w$ with f in $\Sigma \setminus \Omega^+$ or $u \to v$, with u, v Ω^+-terms, and (iii) the rules $\boldsymbol{E} \cup \boldsymbol{G}$ are terminating modulo B. Then $(\Sigma, B, \boldsymbol{E} \cup \boldsymbol{G})$ is admissible and $(\Sigma, E \cup B) \equiv_{sem} (\Sigma, E \cup G \cup B)$.*

The proof of Theorem 6 below can be found in Appendix A.

Theorem 6. (Axiom Internalization). *Let $(\Sigma, E \cup B)$ be an admissible equational program sufficiently complete w.r.t. Ω and with constructor subspecification $(\Omega^+, E_{\Omega^+} \cup B_{\Omega^+})$, and let $B' = B'_\Delta \uplus B'_{\Omega^+}$ be a collection of associative and/or commutative and/or unit axioms, with B'_Δ associative and/or commutative $(A \vee C)$ Δ-axioms (resp. B'_{Ω^+} Ω^+-axioms), both general enough to impose B'_Δ axioms for all binary operators $f \in \Delta$ subsort-overloaded with those appearing in B'_Δ (resp. all binary operators $c \in \Omega^+$ subsort-overloaded with those appearing in B'_{Ω^+}), and making Σ $B \cup B'$-preregular. If $(\Sigma, E \cup B) \models_{ind} B'$ and the rules E are terminating modulo $B \cup B'$, then $(\Sigma, E \cup B \cup B')$ is admissible and $(\Sigma, E \cup B) \equiv_{sem} (\Sigma, E \cup B \cup B')$.*

We can use Theorems 5 and 6 to give, in the theorem below (whose proof can be found in Appendix A), conditions ensuring $\mathbb{C}_{\Omega^+/E_{\Omega^+},B_{\Omega^+}} = \mathbb{C}_{\Omega^+/E'_{\Omega^+},B'_{\Omega^+}}$ for admissible constructor specifications $(\Omega^+, E_{\Omega^+} \cup B_{\Omega^+})$ and $(\Omega^+, E'_{\Omega^+} \cup B'_{\Omega^+})$. This yields a proof method to verify Condition (1) in Theorem 4.

Theorem 7. (Constructor Program Equivalence). *Let $(\Omega^+, E_{\Omega^+} \cup B_{\Omega^+})$ and $(\Omega^+, E'_{\Omega^+} \cup B'_{\Omega^+})$ be admissible constructor specifications such that:*

1. *Ω^+ is $B_{\Omega^+} \cup B'_{\Omega^+}$-preregular and the rules $\boldsymbol{E}_{\Omega^+} \cup \boldsymbol{E}'_{\Omega^+}$ are terminating modulo $B_{\Omega^+} \cup B'_{\Omega^+}$.*
2. *$(\Omega^+, E_{\Omega^+} \cup B_{\Omega^+}) \models_{ind} B'_{\Omega^+} \setminus B_{\Omega^+}$ and $(\Omega^+, E'_{\Omega^+} \cup B'_{\Omega^+}) \models_{ind} B_{\Omega^+} \setminus B'_{\Omega^+}$.*
3. *$(\Omega^+, E_{\Omega^+} \cup B_{\Omega^+} \cup B'_{\Omega^+}) \models_{ind} E'_{\Omega^+} \setminus E_{\Omega^+}$ and $(\Omega^+, E'_{\Omega^+} \cup B_{\Omega^+} \cup B'_{\Omega^+}) \models_{ind} E_{\Omega^+} \setminus E'_{\Omega^+}$.*

Then $\mathbb{C}_{\Omega^+/E_{\Omega^+},B_{\Omega^+}} = \mathbb{C}_{\Omega^+/E'_{\Omega^+},B'_{\Omega^+}}$.

Example 4. (*Example 3 continued*). The only pending issue to prove the equivalence of natural arithmetic programs in Example 3 was the constructor program equivalence NAT+ACU $\equiv_{sem}$ NAT+AU. Since there are no rewrite rules in either NAT+ACU or NAT+AU, NAT+ACU is ACU-preregular, and $AU \setminus ACU = \emptyset$, by Theorem 7, to show NAT+ACU $\equiv_{sem}$ NAT+AU we only need to prove NAT+AU $\models_{ind} N + M = M + N$. The NuITP proof script can be found in Appendix B.

The following theorem, whose proof is given in Appendix A, shows that if $\mathcal{E} \subseteq \mathcal{E}''$ and $\mathcal{E} \equiv_{sem} \mathcal{E}''$, then, any $\mathcal{E}'$ such that $\mathcal{E} \subseteq \mathcal{E}' \subseteq \mathcal{E}''$ is, under mild conditions, admissible and semantically equivalent to both $\mathcal{E}$ and $\mathcal{E}''$.

Theorem 8. (Module Interpolation). *Let $\mathcal{E} = (\Sigma, E \uplus B)$ and $\mathcal{E}'' = (\Sigma, E \uplus E_1 \uplus E_2 \uplus B \uplus B_1 \uplus B_2)$ be admissible theories with constructors Ω such that $\mathcal{E} \equiv_{sem} \mathcal{E}''$. Then, any $\mathcal{E}' = (\Sigma, E \uplus E_1 \uplus B \uplus B_1)$ with $B_1 = B_{1_\Delta} \uplus B_{1_{\Omega^+}}$ associative and/or commutative and/or unit axioms, with B_{1_Δ} $A \vee C$ Δ-axioms (resp. $B_{1_{\Omega^+}}$ Ω^+-axioms), both general enough to impose B_{1_Δ} axioms for all binary operators $f \in \Delta$ subsort-overloaded with those appearing in B_{1_Δ} (resp. all binary operators $c \in \Omega^+$ subsort-overloaded with those appearing in $B_{1_{\Omega^+}}$), is admissible and we have, $\mathcal{E} \equiv_{sem} \mathcal{E}' \equiv_{sem} \mathcal{E}''$.*

Internalize and Conquer. In inductive theorem proving the verification of properties for an equational program $\mathcal{E}$ is greatly eased by using *formula simplification techniques*. The most basic such technique is simplification with $\mathcal{E}$'s equations, and also with *equations already proved as lemmas*. Often, a substantial part of the proof effort is spent proving lemmas that are equations providing essential knowledge needed in proving the main goal. There are, however, several difficulties: first, an equation proved as a lemma may fail to be terminating, which causes looping; second, lemma proving tends to happen in an ad-hoc and by-need fashion, and in a context of current knowledge that may not be optimal, so that lemma proving may require significant effort. The end result in interactive inductive theorem proving can be a tedious proof that is longer than desirable.

These difficulties can be substantially reduced by using Theorems 5 and 6 as a key part of an inductive theorem proving methodology that I call *internalize and conquer*, based on three simple ideas:

1. Think of program properties, not in isolation, but as part of a *cluster* of properties.
2. Arrange the properties to be proved in a "low hanging fruit" manner: from simpler to more complex, in the hope that, as simpler properties get proved, their accumulated knowledge will have a *domino effect* in making the proofs of more complex properties much simpler.
3. Use Theorems 5 and 6 to automatically *internalize* all the knowledge of already-proved properties so that, hopefully, *no lemmas need to be proved*.

Idea (2) is part of the best practice, not just in theorem proving, but in Mathematics, where proofs of complex theorems become trivial if simpler results have been proved beforehand. The above methodology is at the same time a *method to prove program equivalences*, since Theorems 5 and 6 allow us to prove the equivalence between an equational program $\mathcal{E} = (\Sigma, E \cup B)$ and a richer, yet equivalent, program $\mathcal{E}' = (\Sigma, E \cup E' \cup B \cup B')$. What Ideas (1)–(3) help us do is to arrange the properties $B' \cup E'$ to be proved in order of complexity, and use Theorems 5 and 6 to incrementally build a chain of program equivalences that starts in $\mathcal{E}$ and ends in $\mathcal{E}'$. Let us see an example.

Example 5. Consider the following Maude specification of natural number addition, multiplication and exponentiation in Peano notation:

```
fmod NATURAL-ARITH is
  sorts Nat NzNat .  subsort NzNat < Nat .
  op 0 : -> Nat [ctor metadata "1"] .
  op s : Nat -> NzNat [ctor metadata "2"] .
  op _+_ : Nat Nat -> Nat [metadata "3"] .
  op _*_ : Nat Nat -> Nat [metadata "4"] .
  op _*_ : NzNat NzNat -> NzNat [metadata "5"] .
  op _^_ : NzNat Nat -> NzNat [metadata "6"] .
  vars n m k : Nat .  vars n' k' m' : NzNat .
  eq n + 0 = n .
  eq n + s(m) = s(n + m) .
  eq n * 0 = 0 .
  eq n * s(m) = n + (n * m) .
  eq n' ^ 0 = s(0) .
  eq n' ^ s(m) = n' * (n' ^ m) .
endfm
```

where the `metadata` attribute is used to define an RPO order among function symbols. This order can later be used to automatically try to orient equations proved as inductive properties as terminating rewrite rules. A fundamental cluster of properties one would like to prove about `NATURAL-ARITH` is that its canonical term algebra: (i) satisfies the axioms of the theory of *commutative semirings*, i.e., it is a commutative monoid under addition with unit 0, a commutative monoid under multiplication with unit $s(0)$, and multiplication is distributive over addition; and (ii) it also satisfies expected exponentiation properties such as $x^{y+z} = x^y * x^z$, $x^{y*z} = (x^y)^z$, and $(x^y)^z = (x^z)^y$, with x of sort `NzNat` and y, z of sort `Nat`. Following the internalize and conquer methodology, all these properties should be arranged in order of complexity: properties of addition before those of multiplication and these before those of exponentiation. Commutativity properties for $+$ and $*$ are known to be harder to prove. Their proofs can be made considerably simpler if their defining equations, which in `NATURAL-ARITH` recurse on their second argument, are proved equivalent to the analogous equations recursing on the first argument. This suggest the general idea of extending the cluster of properties to be proved by some auxiliary properties that we suspect would otherwise show up as lemmas to be proved. In Maude's **NuITP** inductive theorem prover [3] the internalization associated to Theorem 5 is supported by the `internalize` command, and that associated to Theorem 6 by the `internalize as assoc` and `internalize as comm` commands. The entire proof of all the above properties can be obtained as an interleaving of **NuITP** commands to prove an equation, and `internalize` commands (including those for internalizing associativity or commutativity axioms) in such way that each equational property is proved by a *single* **NuITP** command, without any auxiliary lemmas. The **NuITP** proof script can be found in Appendix B.

Proving Ground Convergence Through Internalization. The internalize and conquer methodology builds a "telescope" of equational programs from $\mathcal{E} = (\Sigma, E \cup B)$ to $\mathcal{E}' = (\Sigma, E \cup E' \cup B \cup B')$,

$$(\Sigma, E \cup B) \subset (\Sigma, E \cup E_1 \cup B \cup B_1) \ldots (\Sigma, E \cup E_{k-1} \cup B \cup B_{k-1}) \subset (\Sigma, E \cup E' \cup B \cup B')$$

while at the same time *proving* that all programs in the telescope (or, by the Module Interpolation Theorem, other programs that could be inserted in the telescope) are indeed programs, i.e., are *ground convergent*. Ground convergence of an equational theory $\mathcal{E} = (\Sigma, E \cup B)$ is an inductive property much harder to prove than convergence, which, assuming termination, is a decidable property for unconditional programs by checking sort-decreasingness and joinability of critical pairs.[6] The reason is that one needs to reason, not about the initial algebra $\mathbb{T}_{\Sigma/E\cup B}$, but about the initial model $\mathbb{T}_{\mathcal{E}}$ of the rewrite theory $\mathcal{E} = (\Sigma, B, E)$, which requires reasoning inductively about the $\to_{E,B}$ and $\downarrow_{E,B}$ relations (see, e.g., [6]). Internalize and conquer is also a new, incremental methodology to prove ground convergence, where the inductive reasoning happens, not on $\mathbb{T}_{\mathcal{E}}$, but on $\mathbb{T}_{\Sigma/E\cup B}$ for inductively equivalent $\mathcal{E}$'s. Let us see an example:

Example 6. Consider the Maude equational program:

```
fmod NATURAL+C*C is
  sorts Nat NzNat .   subsort NzNat < Nat .
  op 0 : -> Nat [ctor metadata "1"] .
  op s : Nat -> NzNat [ctor metadata "2"] .
  op _+_ : Nat Nat -> Nat [comm metadata "3"] .
  op _*_ : Nat Nat -> Nat [comm metadata "4"] .
  op _*_ : NzNat NzNat -> NzNat [comm metadata "5"] .
  vars n m k : Nat .   vars n' k' m' : NzNat .
  eq n + 0 = n .
  eq n + s(m) = s(n + m) .
  eq n * 0 = 0 .
  eq n * s(m) = n + (n * m) .
endfm
```

It is terminating, but its equations are *not* confluent. Maude's Church-Rosser Checker reports the unjoinable critical pair $s(N + (M + (N * M))) = s(M + (N + (N * M)))$. The usual method to prove NATURAL+C*C ground confluent would be to prove that $s(N + (M + (N * M))) \downarrow s(M + (N + (N * M)))$ is an inductive theorem of the rewrite theory $\text{NATURAL} + C * C$ using, for example, the inductive inference system in [6]. The internalize and conquer methodology offers an alternative way to prove NATURAL+C*C ground confluent. This holds because NATURAL-ARITH in Example 5, as shown in its **NuITP** proof script, is semantically equivalent to a supermodule of it which has internalized the associativity and commutativity axioms for + and *. But then, by the Module Interpolation Theorem, NATURAL-ARITH is semantically equivalent to the admissible module:

```
fmod NATURAL-ARITH+C*C is
  sorts Nat NzNat .   subsort NzNat < Nat .
```

---

[6] If B includes axioms for associative but non-commutative symbols, it is possible in theory to have an infinite set of such critical pairs; but in practice this infinity can be avoided in most cases for three reasons: (i) when computing critical pairs the B-unifications involved are *disjoint* ones; (ii) a non-left-linear rule can always be made left-linear by making the non-linearity constraint part of its condition; and (iii) it is well-known that disjoint A-unification of left-linear terms is finitary.

```
  op 0 : -> Nat [ctor metadata "1"] .
  op s : Nat -> NzNat [ctor metadata "2"] .
  op _+_ : Nat Nat -> Nat [comm metadata "3"] .
  op _*_ : Nat Nat -> Nat [comm metadata "4"] .
  op _*_ : NzNat NzNat -> NzNat [comm metadata "5"] .
  op _^_ : NzNat Nat -> NzNat [metadata "6"] .
  vars n m k : Nat .   vars n' k' m' : NzNat .
  eq n + 0 = n .
  eq n + s(m) = s(n + m) .
  eq n * 0 = 0 .
  eq n * s(m) = n + (n * m) .
  eq n' ^ 0 = s(0) .
  eq n' ^ s(m) = n' * (n' ^ m) .
endfm
```

It then follows from Lemma 1 applied to the semantically equivalent programs NATURAL-ARITH and NATURAL-ARITH+C*C with $\Delta_0 = \{+, *\}$ that NATURAL+C*C, is also an admissible module and therefore ground convergent, as desired.

6 Related Work and Conclusions

This work is most closely related to inductive theorem proving methods and tools for OBJ [12], CafeOBJ [9], and Maude [3,15]. In comparison with that work, the checkable conditions for preserving program properties up and/or down program hierarchies, the methods and conditions to prove program equivalence, and the internalization results and methodology, except for the support of internalization in [3], which is based on an earlier, unpublished version of these results, seem to be new. This work is also most closely related to methods and tools to prove order-sorted equational specifications ground convergent [6,20]. In relation to that work, the internalization method to prove order-sorted equational specifications ground convergence is new.

In conclusion, this work has presented new concepts and methods to make the verification of order-sorted equational programs modulo structural axioms more modular, succinct and reusable, including: (i) preservation of program properties up and/or down program hierarchies; (ii) a notion of equational program equivalence and simple conditions for its verification; and (iii) an *internalize and conquer* methodology and conditions to make: (a) proofs of program equivalence simpler and incremental, (b) inductive proofs shorter; and (c) proofs of ground convergence simpler. Future work includes support for conditional rewrite rules, and more general program equivalence notions. Program equivalence requires that the two canonical term algebras are *identical* and therefore have the same *data representation*. This can be generalized to a notion of *program isomorphism* between programs whose algebraic data types that may have different data representations.

Acknowledgments. I thank the reviewers for their excellent comments and suggestions. Work partially supported under NRL contract N0017323C2002.

A Proofs of Theorems and Lemmas

Proof of Theorem 2.

Proof. First of all, note that, since $\mathbb{C}_{\Sigma/E,B}|_{\Omega^+} = \mathbb{C}_{\Omega^+/E_{\Omega^+},B_{\Omega^+}}$ and $\mathbb{C}_{\Sigma'/E',B'}|_{\Omega'^+} = \mathbb{C}_{\Omega'^+/E'_{\Omega'^+},B'_{\Omega'^+}}$, the S-sorted function of the unique Σ-homomorphism $h : \mathbb{C}_{\Sigma/E,B} \to \mathbb{C}_{\Sigma'/E',B'}|_\Sigma$ coincides with the S-sorted function of the unique Ω^+-homomorphism $h : \mathbb{C}_{\Omega^+/E_{\Omega^+},B_{\Omega^+}} \to \mathbb{C}_{\Omega'^+/E'_{\Omega'^+},B'_{\Omega'^+}}|_{\Omega^+}$. Therefore, in each case (1)–(3) we just need to show that, for each sort $s \in S$, the function $h_s : C_{\Omega^+/E_{\Omega^+},B_{\Omega^+},s} \to C_{\Omega'^+/E'_{\Omega'^+},B'_{\Omega'^+},s}$ is injective, resp. surjective, resp. bijective. Furthermore, since the unique surjective Ω^+-homomorphism $\mathbb{T}_{\Omega^+} \to \mathbb{C}_{\Omega^+/E_{\Omega^+},B_{\Omega^+}}$ maps each $u \in T_{\Omega^+,s}$ to $[u!_{E_{\Omega^+},B_{\Omega^+}}] \in C_{\Omega^+/E_{\Omega^+},B_{\Omega^+},s}$, and the unique Ω^+-homomorphism $\mathbb{T}_{\Omega^+} \to \mathbb{C}_{\Omega'^+/E'_{\Omega'^+},B'_{\Omega'^+}}|_{\Omega^+}$ maps each $u \in T_{\Omega^+,s}$ to $[u!_{E'_{\Omega'^+},B'_{\Omega'^+}}] \in C_{\Omega'^+/E'_{\Omega'^+},B'_{\Omega'^+},s}$, by initiality of $\mathbb{T}_{\Omega^+}$ h_s must be the function $h_s : C_{\Omega^+/E_{\Omega^+},B_{\Omega^+},s} \ni [u!_{E_{\Omega^+},B_{\Omega^+}}]_{B_{\Omega^+}} \mapsto [u!_{E'_{\Omega'^+},B'_{\Omega'^+}}]_{B'_{\Omega'^+}} \in C_{\Omega'^+/E'_{\Omega'^+},B'_{\Omega'^+},s}$.

If the inclusion is **extending** we need to show that for any $[u]_{B_{\Omega^+}}, [v]_{B_{\Omega^+}} \in C_{\Omega^+/E_{\Omega^+},B_{\Omega^+},s}$, if $[u!_{E'_{\Omega'^+},B'_{\Omega'^+}}]_{B'_{\Omega'^+}} = [v!_{E'_{\Omega'^+},B'_{\Omega'^+}}]_{B'_{\Omega'^+}}$, then $[u]_{B_{\Omega^+}} = [v]_{B_{\Omega^+}}$. But $[u!_{E'_{\Omega'^+},B'_{\Omega'^+}}]_{B'_{\Omega'^+}} = [v!_{E'_{\Omega'^+},B'_{\Omega'^+}}]_{B'_{\Omega'^+}}$ just means $u!_{E'_{\Omega'^+},B'_{\Omega'^+}} =_{B'_{\Omega'^+}} v!_{E'_{\Omega'^+},B'_{\Omega'^+}}$. First of all note that, $u!_{E'_{\Omega'^+},B'_{\Omega'^+}} = u$ and $v!_{E'_{\Omega'^+},B'_{\Omega'^+}} = v$, i.e., both terms are in $E'_{\Omega'^+}, B'_{\Omega'^+}$-canonical form. Let me show this for $u!_{E'_{\Omega'^+},B'_{\Omega'^+}}$ since the argument is identical for $v!_{E'_{\Omega'^+},B'_{\Omega'^+}}$. Suppose $u \to_{E'_{\Omega'^+},B'_{\Omega'^+}} w$. This happens iff there is a term w' such that $u!_{U_{\Omega'^+},B'_{A\vee C,\Omega'^+}} = u'$, $u' \to_{E'_{\Omega'^+}\cup U_{\Omega'^+},B'_{A\vee C,\Omega'^+}} w'$ and $w =_{B'_{\Omega'^+}} w'$. But note that $u' = u!_{U_{\Omega'^+},B'_{A\vee C,\Omega'^+}}$ is actually $u' = u!_{U_{\Omega^+},B_{A\vee C,\Omega^+}}$. This is because, (a) u is an Ω^+-term and by assumption (i), if w is an Ω^+-term, then $w =_{B'_{A\vee C,\Omega'^+}} w'$ iff w' is an Ω^+-term and $w =_{B_{A\vee C,\Omega^+}} w'$, and (b), also by assumption (i), for any Ω^+-term u'' if $u'' \to_{U_{\Omega'^+},B'_{A\vee C,\Omega'^+}} u'''$ the rule $f(x,e) \to x$, (resp. $f(e,x) \to x$) applied to u'' to get u''' can always be chosen to be a rule in U_{Ω^+}. This means that u' is an Ω^+-term and $u =_{B_{\Omega^+}} u'$. But since $[u] \in C_{\Omega^+/E_{\Omega^+},B_{\Omega^+},s}$, u' must be in $E_{\Omega^+}, B_{\Omega^+}$-canonical form. This therefore means that a rewrite $u' \to_{E'_{\Omega'^+}\cup U_{\Omega'^+},B'_{A\vee C,\Omega'^+}} w'$ is impossible, since by (a) above and the assumption in (1) such a rewrite must be of the form $u' \to_{E_{\Omega^+}\cup U_{\Omega^+},B_{A\vee C,\Omega^+}} w'$, which is impossible by u' in $E_{\Omega^+}, B_{\Omega^+}$-canonical form. This shows that $u!_{E'_{\Omega'^+},B'_{\Omega'^+}} = u$ and $v!_{E'_{\Omega'^+},B'_{\Omega'^+}} = v$. We furthermore have $u!_{E'_{\Omega'^+},B'_{\Omega'^+}} =_{B'_{\Omega'^+}} v!_{E'_{\Omega'^+},B'_{\Omega'^+}}$, i.e., $u =_{B'_{\Omega'^+}} v$, which holds iff $u!_{U_{\Omega'^+},B'_{A\vee C,\Omega'^+}} =_{B'_{A\vee C,\Omega'^+}} v!_{U_{\Omega'^+},B'_{A\vee C,\Omega'^+}}$, which, reasoning as above, holds iff $u!_{U_{\Omega^+},B_{A\vee C,\Omega^+}} =_{B_{A\vee C,\Omega^+}} v!_{U_{\Omega^+},B_{A\vee C,\Omega^+}}$, which is equivalent to $u =_{B_{\Omega^+}} v$, proving the injectivity of h_s.

If the inclusion is **generated-by** we need to show h surjective. Since $\mathbb{C}_{\Omega'^+/E'_{\Omega'^+},B'_{\Omega'^+}}$ and $\mathbb{C}_{\Omega^+/E_{\Omega^+},B_{\Omega^+}}$ are initial Ω'^+-, resp. Ω^+-, algebras, we have surjective Ω'^+-, resp. Ω^+-, homomorphisms $q' : \mathbb{T}_{\Omega'^+} \to \mathbb{C}_{\Omega'^+/E'_{\Omega'^+}, B'_{\Omega'^+}}$ and $q : \mathbb{T}_{\Omega^+} \to \mathbb{C}_{\Omega^+/E_{\Omega^+}, B_{\Omega^+}}$. But conditions (a)–(b) in (2) mean that $\mathbb{T}_{\Omega'^+}|_{\Omega^+} = \mathbb{T}_{\Omega^+}$, which forces the unique Ω^+-homomorphism $\mathbb{T}_{\Omega^+} \to \mathbb{C}_{\Omega'^+/E'_{\Omega'^+}, B'_{\Omega'^+}}|_{\Omega^+}$ to be $q'|_\Omega$ and therefore surjective, and, by initiality of $\mathbb{T}_{\Omega^+}$, the homomorphism identity $q; h = q'|_{\Omega^+}$, which by $q'|_{\Omega^+}$ surjective forces h to be surjective.

If the inclusion is **protecting**, h is injective by the requiements in (1) and surjective by those in (2), and therefore bijective. □.

Proof of Theorem 3.

Proof. To see the ($\Rightarrow$) implication, since $\mathbb{T}_{\Sigma/E \cup G} \models E$ we have a unique Σ-homomorphism $h : \mathbb{T}_{\Sigma/E} \to \mathbb{T}_{\Sigma/E \cup G}$. And since $\mathbb{T}_{\Sigma/E} \models E \cup G$, we also have a unique Σ-homomorphism $g : \mathbb{T}_{\Sigma/E \cup G} \to \mathbb{T}_{\Sigma/E}$. But then, the initiality of $\mathbb{T}_{\Sigma/E}$ forces $h; g = id_{\mathbb{T}_{\Sigma/E}}$, and the initiality of $\mathbb{T}_{\Sigma/E \cup G}$ forces $g; h = id_{\mathbb{T}_{\Sigma/E \cup G}}$. Therefore, we have an isomorphism: $\mathbb{T}_{\Sigma/E} \cong \mathbb{T}_{\Sigma/E \cup G}$. We will be done of we prove the following lemma:

Lemma 2. *Let E, E' be two sets of Σ-equations such that $\mathbb{T}_{\Sigma/E} \cong \mathbb{T}_{\Sigma/E'}$. Then, $\mathbb{T}_{\Sigma/E} = \mathbb{T}_{\Sigma/E'}$.*

Proof. $\mathbb{T}_{\Sigma/E}$ and $\mathbb{T}_{\Sigma/E'}$ are uniquely determined by the respective *ground* equality relations $=_E \cap T_\Sigma^2$ and $=_{E'} \cap T_\Sigma^2$. We just need to show $(=_E \cap T_\Sigma^2) = (=_{E'} \cap T_\Sigma^2)$. Since we have a Σ-isomorphism $h : \mathbb{T}_{\Sigma/E} \to \mathbb{T}_{\Sigma/E'}$, and unique Σ-homomorphisms $[\_]_E : \mathbb{T}_\Sigma \to \mathbb{T}_{\Sigma/E}$, and $[\_]_{E'} : \mathbb{T}_\Sigma \to \mathbb{T}_{\Sigma/E}$, the initiality of $\mathbb{T}_\Sigma$ forces $[\_]_E; h = [\_]_{E'}$, i.e., $h_s([t]_E) = [t]_{E'}$ for each $t \in T_{\Sigma,s}, s \in S$. Let $t \in T_{\Sigma,s}$ and $t' \in T_{\Sigma,s'}$ with $t =_E t'$. Then $[s] = [s']$ and, by h order-sorted Σ-homomorphism and $[t]_E = [t']_E$, we must have $h_s([t]_E) = h_{s'}([t']_E)$, which forces:

$$h_s([t]_E) = [t]_{E'} = [t']_{E'} = h_{s'}([t']_E)$$

giving us the containment $(=_E \cap T_\Sigma^2) \subseteq (=_{E'} \cap T_\Sigma^2)$. Using the inverse isomorphism h^{-1} we likewise get $(=_{E'} \cap T_\Sigma^2) \subseteq (=_E \cap T_\Sigma^2)$, giving us $(=_E \cap T_\Sigma^2) = (=_{E'} \cap T_\Sigma^2)$, as desired. □

To see the ($\Leftarrow$) implication, since $\mathbb{T}_{\Sigma/E} = \mathbb{T}_{\Sigma/E \cup G}$ we have $\mathbb{T}_{\Sigma/E} \models G$, which exactly means $(\Sigma, E) \models_{ind} G$. □

Proof of Lemma 1

Proof. It follows from the assumptions on E_{Δ_0} (resp. E'_{Δ_0}) and B_Δ, that for any $t \in T_{\Delta_0 \uplus \Omega^+}$, if $t \to_{E,B} t'$ (resp. $t \to_{E',B'} t'$), then $t' \in T_{\Delta_0 \uplus \Omega^+}$ and $t \to_{E_{\Delta_0} \cup E_{\Omega_+}, B_{\Delta_0} \cup B_{\Omega_+}} t'$ (resp. $t \to_{E'_{\Delta_0} \cup E'_{\Omega_+}, B'_{\Delta_0} \cup B'_{\Omega_+}} t'$). Therefore, both $(\Delta_0 \uplus \Omega^+, E_{\Delta_0} \cup E_{\Omega_+} \cup B_{\Delta_0} \cup B_{\Omega^+})$ and $(\Delta_0 \uplus \Omega^+, E'_{\Delta_0} \cup E'_{\Omega_+} \cup B'_{\Delta_0} \cup B'_{\Omega^+})$ are sufficiently complete w.r.t. Ω, ground convergent, and therefore admissible. Furthermore, for each $t \in T_{\Delta_0 \uplus \Omega^+}$, $t!_{E,B} = t!_{E_{\Delta_0} \cup E_{\Omega_+}, B_{\Delta_0} \cup B_{\Omega_+}}$ and $t!_{E',B'} = t!_{E'_{\Delta_0} \cup E'_{\Omega_+}, B'_{\Delta_0} \cup B'_{\Omega_+}}$, which forces

$\mathbb{C}_{\Delta \uplus \Omega^+/E,B}|_{\Delta_0 \uplus \Omega^+} = \mathbb{C}_{\Delta \uplus \Omega^+/E_{\Delta_0} \cup E_{\Omega_+}, B_{\Delta_0} \cup B_{\Omega^+}}$ and $\mathbb{C}_{\Delta \uplus \Omega^+/E',B'}|_{\Delta_0 \uplus \Omega^+} = \mathbb{C}_{\Delta \uplus \Omega^+/E'_{\Delta_0} \cup E'_{\Omega_+}, B'_{\Delta_0} \cup B'_{\Omega^+}}$, which, by $\mathcal{E} \equiv_{sem} \mathcal{E}'$, forces $(\Delta_0 \uplus \Omega^+, E_{\Delta_0} \cup E_{\Omega_+} \cup B_{\Delta_0} \cup B_{\Omega^+}) \equiv_{sem} (\Delta_0 \uplus \Omega^+, E'_{\Delta_0} \cup E'_{\Omega_+} \cup B'_{\Delta_0} \cup B'_{\Omega^+})$. □

Proof of Theorem 4.

Proof. To see ($\Rightarrow$), note that semantic equivalence forces (1), since $\mathbb{C}_{\Sigma/E,B} = \mathbb{C}_{\Sigma/E',B'}$ implies that $\mathbb{C}_{\Sigma/E,B}|_\Omega = \mathbb{C}_{\Omega/E_\Omega, B_\Omega} = \mathbb{C}_{\Omega/E'_\Omega, B'_\Omega} = \mathbb{C}_{\Sigma/E',B'}|_\Omega$; and also forces (2), since $\mathcal{E} \equiv_{ind} \mathcal{E}'$ means that $\mathbb{T}_{\Sigma/E \cup B} = \mathbb{T}_{\Sigma/E' \cup B'}$, which forces $(\Sigma, E \cup B) \models_{ind} (E'_\Delta \setminus E_\Delta) \cup (B'_\Delta \setminus B_\Delta)$.

To prove the ($\Leftarrow$) implication we first prove:

Lemma 3. *For any two canonical term algebras $\mathbb{C}_{\Sigma/E,B}$ and $\mathbb{C}_{\Sigma/E',B'}$ with respective constructor subspecifications $(\Omega, E_\Omega \cup B_\Omega)$ and $(\Omega, E'_\Omega \cup B'_\Omega)$ and such that $\mathbb{C}_{\Omega/E_\Omega, B_\Omega} = \mathbb{C}_{\Omega/E'_\Omega, B'_\Omega}$, $\mathbb{C}_{\Sigma/E,B} = \mathbb{C}_{\Sigma/E',B'}$ iff for each Σ-term t, $t!_{E/B} =_{B_\Omega} t!_{E'/B'}$.*

Proof. To see the ($\Rightarrow$) implication, note that the unique Σ-homomorphism $\mathbb{T}_\Sigma \to \mathbb{C}_{\Sigma/E,B}$ maps each $t \in \mathbb{T}_\Sigma$ to $[t!_{E/B}]_{B_\Omega} \in \mathbb{C}_{\Omega/E_\Omega, B_\Omega}$. Therefore, $\mathbb{C}_{\Sigma/E,B} = \mathbb{C}_{\Sigma/E',B'}$ means that for each $t \in \mathbb{T}_\Sigma$, $[t!_{E/B}]_{B_\Omega} = [t!_{E'/B'}]_{B_\Omega}$, and therefore that $t!_{E/B} =_{B_\Omega} t!_{E'/B'}$.

To see the ($\Leftarrow$) implication, since $\mathbb{C}_{\Omega/E_\Omega,B_\Omega} = \mathbb{C}_{\Omega/E'_\Omega,B'_\Omega}$, $\mathbb{C}_{\Sigma/E,B}$ and $\mathbb{C}_{\Sigma/E',B'}$ have the same underlying S-sorted set of data elements, and the same interpretation $c_{\mathbb{C}_{\Sigma/E,B}} = c_{\mathbb{C}_{\Sigma/E',B'}}$ for each constructor operator $c \in \Omega$. Therefore, to show $\mathbb{C}_{\Sigma/E,B} = \mathbb{C}_{\Sigma/E',B'}$ we only need to show that for each $f \in \Delta$ we have $f_{\mathbb{C}_{\Sigma/E,B}} = f_{\mathbb{C}_{\Sigma/E',B'}}$. Indeed, $f_{\mathbb{C}_{\Sigma/E,B}}([u_1], \ldots, [u_n]) = [f(u_1, \ldots, u_n)!_{E/B}] = [f(u_1, \ldots, u_n)!_{E'/B'}] = f_{\mathbb{C}_{\Sigma/E',B'}}([u_1], \ldots, [u_n])$. □

Now note that (1) (using Theorems 3 and 1) and (2) force $(\Sigma, E \cup B) \models_{ind} E' \cup B'$, which forces $t!_{E/B} =_{E \cup B} t!_{E'/B'}$, which, by the Church-Rosser property, then forces $t!_{E/B} =_{B_\Omega} (t!_{E'/B'})!_{E/B}$, which by (1) and Lemma 3 forces $\mathbb{C}_{\Sigma/E,B} = \mathbb{C}_{\Sigma/E',B'}$. □

Proof of Theorem 5.

Proof. $(\Sigma, B, E \cup G)$ will be admissible if we prove that $(\Sigma, B, E \cup G)$ is locally ground confluent modulo B. Let $t, u, v \in \mathbb{T}_\Sigma$ be such that $u \;_{E \cup G/B} \leftarrow t \to_{E \cup G/B} v$. We need to show that $u \downarrow_{E \cup G/B} v$. This will hold if we prove that $u \downarrow_{E/B} v$. But since $(\Sigma, E \cup B) \models_{ind} G$, by Theorem 3 this forces $u =_{E \cup B} v$, which, since E is gound convergent modulo B, forces $u \downarrow_{E/B} v$. Next we have to show that $\mathbb{C}_{\Sigma/E,B} = \mathbb{C}_{\Sigma/E \cup G, B}$. Let us first prove that for each $t \in \mathbb{T}_\Sigma$ $t!_{E,B} =_{B_\Omega} t!_{E \cup G, B}$. Since $t!_{E,B} =_{E \cup G \cup B} t!_{E \cup G, B}$, by $(\Sigma, E \cup B) \models_{ind} G$ and Theorem 3, $t!_{E,B} =_{E \cup B} t!_{E \cup G, B}$, which by the ground Church-Rosser Theorem holds iff $t!_{E,B} =_{B_\Omega} (t!_{E \cup G, B})!_{E,B}$. But, by definition, $t!_{E \cup G, B}$ is already in E, B-canonical form, so that $t!_{E,B} =_{B_\Omega} t!_{E \cup G, B}$. This means that

$\mathbb{C}_{\Sigma/E,B}$ and $\mathbb{C}_{\Sigma/E\cup G,B}$ have the same S-sorted set of data elements. So, to prove $\mathbb{C}_{\Sigma/E,B} = \mathbb{C}_{\Sigma/E\cup G,B}$ it is enough to prove that for each $f \in \Sigma$ we have the function identity $f_{\mathbb{C}_{\Sigma/E,B}} = f_{\mathbb{C}_{\Sigma/E\cup G,B}}$. But this follows easily from $t!_{E,B} =_{B_\Omega} t!_{E\cup G,B}$, as in the proof of Lemma 3. □

Proof of Theorem 6.

Proof. To show $(\Sigma, E \cup B \cup B')$ admissible we need to show that the rules E are locally ground confluent modulo $B \cup B'$. Let $t, u, v \in T_\Sigma$ be such that $u \;{}_{E/B\cup B'}\!\!\leftarrow t \rightarrow_{E/B\cup B'} v$. We need to show that $u \downarrow_{E/B\cup B'} v$. This will hold if we prove $u \downarrow_{E/B} v$. But since $(\Sigma, E\cup B) \models_{ind} B'$, Theorem 3 forces $u =_{E\cup B} v$, which, since E is ground confluent modulo B, forces $u \downarrow_{E/B} v$, as desired.

We will be done if we show that $\mathbb{C}_{\Sigma/E,B} = \mathbb{C}_{\Sigma/E,B\cup B'}$. First of all note that for any ground Σ-term $t \in T_\Sigma$, $t!_{E,B} =_{B_\Omega} t!_{E,B\cup B'}$. This is because, by Theorem 3, $t!_{E,B} =_{E\cup B} t_{E,B\cup B'}$, by the ground Church Rosser Theorem $t!_{E,B} =_{B_\Omega} (t!_{E,B\cup B'})!_{E,B}$ and by $t!_{E,B\cup B'}$ already being in E, B-normal form $t!_{E,B} =_{B_\Omega} t!_{E,B\cup B'}$. Second, $\mathbb{C}_{\Sigma/E,B}$ and $\mathbb{C}_{\Sigma/E,B\cup B'}$ will have the same undelying S-sorted set of data elements if we show that $t!_{E,B} =_{B_\Omega} t!_{E,B\cup B'} \Leftrightarrow t!_{E,B} =_{B_\Omega \cup B'_\Omega} t!_{E,B\cup B'}$, since this shows that $[t!_{E,B}]_{B_\Omega} = [t!_{E,B\cup B'}]_{B_\Omega \cup B'_\Omega}$. This only requires showing $t!_{E,B} =_{B_\Omega \cup B'_\Omega} t!_{E,B\cup B'} \Rightarrow t!_{E,B} =_{B_\Omega} t!_{E,B\cup B'}$. But, by Theorem 3, $t!_{E,B} =_{B_\Omega \cup B'_\Omega} t!_{E,B\cup B'} \Rightarrow t!_{E,B} =_{E\cup B} t!_{E,B\cup B'}$, which, as shown above, forces $t!_{E,B} =_{B_\Omega} t!_{E,B\cup B'}$, as desired. The only remaining task is to show that for each $f \in \Sigma$, $f_{\mathbb{C}_{\Sigma/E,B}} = f_{\mathbb{C}_{\Sigma/E,B\cup B'}}$. But this follows easily from $t!_{E,B} =_{B_\Omega} t!_{E,B\cup B'}$, exactly as in the proof of Lemma 3. □

Proof of Theorem 7

Proof. Assuming (1), (2) and Theorem 6 (with $\Delta = \emptyset$) yields $\mathbb{C}_{\Omega/E_\Omega, B_\Omega} = \mathbb{C}_{\Omega/E_\Omega, B_\Omega \cup B'_\Omega}$ and $\mathbb{C}_{\Omega/E'_\Omega, B'_\Omega} = \mathbb{C}_{\Omega/E'_\Omega, B_\Omega \cup B'_\Omega}$. Assuming (1), (3) and Theorem 5 yields $\mathbb{C}_{\Omega/E_\Omega, B_\Omega \cup B'_\Omega} = \mathbb{C}_{\Omega/E_\Omega \cup E'_\Omega, B_\Omega \cup B'_\Omega} = \mathbb{C}_{\Omega/E'_\Omega, B_\Omega \cup B'_\Omega}$. Therefore,

$$\mathbb{C}_{\Omega/E_\Omega, B_\Omega} = \mathbb{C}_{\Omega/E_\Omega, B_\Omega \cup B'_\Omega} = \mathbb{C}_{\Omega/E_\Omega \cup E'_\Omega, B_\Omega \cup B'_\Omega} = \mathbb{C}_{\Omega/E'_\Omega, B_\Omega \cup B'_\Omega} = \mathbb{C}_{\Omega/E'_\Omega, B'_\Omega}$$

as desired. □

Proof of Theorem 8

Proof. Since $\mathcal{E} \equiv_{sem} \mathcal{E}''$ whe have $\mathcal{E} \models_{ind} E_1 \cup B_1$. Since by hypothesis the rules $E \cup E_1$ are terminating modulo $B \cup B_1 \cup B_2$, they are a fortiori terminating modulo $B \cup B_1$. Therefore, Theorem 5 applies and we have $(\Sigma, E \cup E_1 \cup B)$ admissible and $\mathcal{E} \equiv_{sem} (\Sigma, E \cup E_1 \cup B)$, which implies $(\Sigma, E\cup E_1 \cup B) \models_{ind} B_1$, and, since Σ is $B \cup B_1 \cup B_2$-preregular, a fortiori Σ is $B \cup B_1$-preregular, so that Theorem 5 applies and we have $\mathcal{E}'$ admissible and $(\Sigma, E\cup E_1 \cup B) \equiv_{sem} \mathcal{E}'$. Therefore, by symmetry and transitivity of $\equiv_{sem}$, we get $\mathcal{E} \equiv_{sem} \mathcal{E}' \equiv_{sem} \mathcal{E}''$. □

B NuITP Proof Scripts

To prove inductive theorems with the **NuITP** all modules should previously be entered in Maude after giving the Maude command `set include BOOL off` . This is because the `BOOL` module, which has several built-in features, would otherwise be added by default; but the **NuITP** does not expect any built-in features in the modules it proves properties about.

NuITP Proof Script for Example 3

```
set module NAT+ACU*AC .

set goal  ((K:Natural * (N':NzNatural + 1)) =
                        (K:Natural + (K:Natural * N':NzNatural))) .

apply eps to 0 .
```

NuITP Proof Script for Example 4

```
set module NAT+ACU .

genset ACUG for Natural is 0 ;; (1 + N:NzNatural) .

set goal ((N:Natural + M:Natural) = (M:Natural + N:Natural)) .

apply gsi* to 0 on $1:Natural .
```

NuITP Proof Script for Example 5

```
set module NATURAL-ARITH .

genset SIND for Nat is 0 ;; s(N:Nat) .

set goal ((0 + Y:Nat = Y:Nat) /\ (s(X:Nat) + Y:Nat) = s(X:Nat + Y:Nat)) .

apply gsi! to 0 on $2:Nat .

internalize .

set goal X:Nat + (Y:Nat + Z:Nat) = (X:Nat + Y:Nat) + Z:Nat .

apply gsi! to 0 on $3:Nat .

internalize as assoc .

set goal (X:Nat + Y:Nat = Y:Nat + X:Nat) .

apply gsi! to 0 on $1:Nat .

internalize as comm .
```

```
set goal X:Nat * (Y:Nat + Z:Nat) = (X:Nat * Y:Nat) + (X:Nat * Z:Nat) .

apply gsi! to 0 on $2:Nat .

internalize .

set goal ((0 * Y:Nat = 0) /\ (s(X:Nat) * Y:Nat) = (Y:Nat + (X:Nat * Y:Nat))) .

apply gsi! to 0 on $2:Nat .

internalize .

set goal (Y:Nat + Z:Nat) * X:Nat = (Y:Nat * X:Nat) + (Z:Nat * X:Nat) .

apply gsi! to 0 on $2:Nat .

internalize .

set goal X:Nat * (Y:Nat * Z:Nat) = (X:Nat * Y:Nat) * Z:Nat .

apply gsi! to 0 on $3:Nat .

internalize as assoc .

set goal (X:Nat * Y:Nat = Y:Nat * X:Nat) .

apply gsi! to 0 on $2:Nat .

internalize as comm .


set goal (X:NzNat ^ (Y:Nat + Z:Nat) = (X:NzNat ^ Y:Nat) * (X:NzNat ^ Z:Nat)) .

apply gsi! to 0 on $2:Nat .

internalize .

set goal (X:NzNat ^ (Y:Nat * Z:Nat) = (X:NzNat ^ Y:Nat) ^ Z:Nat) .

apply gsi! to 0 on $3:Nat .

internalize .

set goal (X:NzNat ^ Y:Nat) ^ Z:Nat = (X:NzNat ^ Z:Nat) ^ Y:Nat .

apply eps to 0 .

internalize .
```

References

1. Clavel, M., Meseguer, J., Palomino, M.: Reflection in membership equational logic, many-sorted equational logic, Horn logic with equality, and rewriting logic. Theor. Comput. Sci. **373**, 70–91 (2007)
2. Dershowitz, N., Jouannaud, J.P.: Rewrite systems. In: van Leeuwen, J. (ed.) Handbook of Theoretical Computer Science, vol. B, pp. 243–320. North-Holland (1990)
3. Durán, F., Escobar, S., Meseguer, J., Sapiña, J.: NuITP alpha 21—an inductive theorem prover for maude equational theories. Available at https://nuitp.webs.upv.es/
4. Durán, F., Eker, S., Escobar, S., Martí-Oliet, N., Meseguer, J., Rubio, R., Talcott, C.L.: Programming and symbolic computation in Maude. J. Log. Algebraic Methods Program. **110** (2020)
5. Durán, F., Lucas, S., Meseguer, J.: Termination modulo combinations of equational theories. In: Frontiers of Combining Systems, 7th International Symposium, FroCoS 2009, Trento, Italy, September 16-18, 2009. Proceedings. Lecture Notes in Computer Science, vol. 5749, pp. 246–262. Springer, Berlin (2009)
6. Durán, F., Meseguer, J., Rocha, C.: Ground confluence of order-sorted conditional specifications modulo axioms. J. Log. Algebraic Methods Program. **111**, 100513 (2020)
7. Escobar, S., Sasse, R., Meseguer, J.: Folding variant narrowing and optimal variant termination. J. Algebraic Logic Program. **81**, 898–928 (2012)
8. Futatsugi, K., Diaconescu, R.: CafeOBJ Report. World Scientific (1998)
9. Futatsugi, K.: Advances of proof scores in CafeOBJ. Sci. Comput. Program. **224**, 102893 (2022). https://doi.org/10.1016/j.scico.2022.102893
10. Goguen, J., Meseguer, J.: Order-sorted algebra I: Equational deduction for multiple inheritance, overloading, exceptions and partial operations. Theor. Comput. Sci. **105**, 217–273 (1992)
11. Goguen, J., Winkler, T., Meseguer, J., Futatsugi, K., Jouannaud, J.P.: Introducing OBJ. In: Software Engineering with OBJ: Algebraic Specification in Action, pp. 3–167. Kluwer (2000)
12. Goguen, J.A.: Theorem proving and algebra. CoRR **abs/2101.02690** (2021). https://arxiv.org/abs/2101.02690
13. Hodges, W.: A Shorter Model Theory. Cambridge, UP (1997)
14. Lucas, S., Meseguer, J.: Normal forms and normal theories in conditional rewriting. J. Log. Algebr. Meth. Program. **85**(1), 67–97 (2016)
15. Meseguer, J., Skeirik, S.: Inductive reasoning with equality predicates, contextual rewriting and variant-based simplification. In: Proceedings of WRLA 2020. LNCS, vol. 12328, pp. 114–135. Springer, Berlin (2020)
16. Meseguer, J.: Conditional rewriting logic as a unified model of concurrency. Theor. Comput. Sci. **96**(1), 73–155 (1992)
17. Meseguer, J.: Membership algebra as a logical framework for equational specification. In: Proceedings of WADT'97, pp. 18–61. Springer LNCS 1376 (1998)
18. Meseguer, J.: Variant-based satisfiability in initial algebras. Sci. Comput. Program. **154**, 3–41 (2018)
19. Meseguer, J., Goguen, J.: Initiality, induction and computability. In: Nivat, M., Reynolds, J. (eds.) Algebraic Methods in Semantics, pp. 459–541. Cambridge University Press (1985)
20. Meseguer, J., Skeirik, S.: On ground convergence and completeness of conditional equational program hierarchies. In: Rewriting Logic and Its Applications—14th

International Workshop, WRLA@ETAPS 2022, Munich, Germany, April 2-3, 2022. Lecture Notes in Computer Science, vol. 13252, pp. 191–211. Springer, Berlin (2022)

Equivalence Checking of Quantum Circuits Based on Dirac Notation in Maude

Canh Minh Do[✉] and Kazuhiro Ogata

Japan Advanced Institute of Science and Technology (JAIST), Nomi, Japan
{canhdo,ogata}@jaist.ac.jp

Abstract. This paper presents an approach to checking the equivalence of quantum circuits based on Dirac notation in Maude. Specifically, we specify quantum states and quantum gates in Dirac notation with scalars and use a set of laws from quantum mechanics and matrix operations to reason about quantum computation. The equivalence of quantum circuits can be reduced to matrix equivalence modulo a global phase in Dirac notation. To achieve this, we compare each column vector of two matrices with respect to the same global phase, making it faster than the actual matrix equivalence check, especially in cases of non-equivalent quantum circuits. Furthermore, our approach enhances the reliability in determining the equivalence problem by taking into account constant inputs for quantum circuits, which have been ignored by state-of-the-art tools. We use Maude, a high-level specification/programming language based on rewriting logic, to develop a support tool called |QCEC⟩ for our approach. Several case studies have been conducted with the tool. These demonstrate the effectiveness of our approach and |QCEC⟩ for the equivalence checking of quantum circuits.

Keywords: Equivalence checking · Quantum circuits · Dirac notation · Maude

1 Introduction

Quantum computing is a rapidly emerging technology that uses the principles of quantum mechanics to solve complex problems beyond the capabilities of current classical computing. Several quantum algorithms have been proposed, showing significant improvements over classical algorithms, such as Shor's fast algorithms for discrete logarithms and factoring in 1994 [20]. Although practical quantum computers capable of running such algorithms effectively are not yet available, recent exponential investments from big companies like IBM, Google, Microsoft, and Intel bring the future of the quantum era within closer reach.

The research was supported by JAIST Research Grant for Fundamental Research and by JSPS KAKENHI Grant Numbers JP23K28060, JP23K19959, JP24K20757.

© The Author(s), under exclusive license to Springer Nature Switzerland AG 2024
K. Ogata and N. Martí-Oliet (Eds.): WRLA 2024, LNCS 14953, pp. 84–103, 2024.
https://doi.org/10.1007/978-3-031-65941-6_5

Quantum circuits are a natural model of quantum computation, comprising qubits and quantum operations (e.g., quantum gates), that can be used to design and implement quantum algorithms. However, quantum circuits are typically used to design quantum algorithms at a high abstraction level without considering specific hardware restrictions. To execute the quantum circuits on an actual quantum device, they have to undergo a *compilation* process, transforming the high abstraction level to a low abstraction level that conforms to all restrictions imposed on the targeted device. More precisely, this compilation process has several key aspects as follows. Firstly, quantum devices natively support only a limited set of quantum operations. Consequently, quantum circuits intended for the target device must be expressed using only these native quantum operations. This requires a *decomposition* (or *translation*) step of non-native quantum operations into sequences of native ones [1,10,12]. Secondly, logical qubits used in quantum circuits have to be mapped to physical qubits on the target device. However, this mapping cannot be arbitrary because the target device imposes restrictions on which physical qubits can interact with each other. To achieve this, a *mapping* (or *routing*) step is required, which involves adding SWAP and Hadmard gates to quantum circuits [5,21,23,25]. Lastly, after the decomposition and mapping steps, the size of quantum circuits tends to increase, posing challenges for their execution on quantum devices due to noise and decoherence effects. Therefore, an *optimization* step is required to reduce the size of quantum circuits in terms of the number of quantum gates [11,14,18]. As a result of these processes, the quantum circuit defined at a high abstraction level and its compiled counterpart defined at a low abstraction level are significantly different. Therefore, it is crucial to verify the equivalence of two quantum circuits based on their functionality.

There are two main approaches to equivalence checking of quantum circuits: one based on quantum decision diagrams [3] and the other based on the ZX calculus [17]. In this work, we propose a different approach to the equivalence checking of quantum circuits based on Dirac notation [6]. Specifically, we specify quantum states (qubits) and quantum gates in Dirac notation with scalars and use a set of laws from quantum mechanics and matrix operations to reason about quantum computation. The functionality of quantum circuits can be described by a sequence of quantum gates, which are represented by unitary matrices. Given two quantum circuits in the form of $U = U_m \ldots U_0$ and $U' = U'_{m'} \ldots U'_0$, the two quantum circuits are considered equivalent if U is equal to U' modulo a global phase, which is physically unobservable [15]. Although Dirac notation provides a canonical form for quantum gates and quantum states, directly comparing U and U' is inefficient because it requires costly matrix-matrix multiplications $U_m \ldots U_0$ and $U'_{m'} \ldots U'_0$ to obtain U and U' for comparison. Moreover, if U is significantly different from U', constructing the entire elements of both matrices is unnecessary. Instead, we can compare each column of two matrices. In other words, for each basis vector $|\phi_i\rangle$ in an orthonormal basis of a Hilbert space, if $U|\phi_i\rangle$ is equal to $U'|\phi_i\rangle$ modulo the same global phase, then the two quantum circuits are equivalent. It is important to note that we may need a few iterations to check the equivalence of non-equivalent quantum circuits, making it faster than the actual matrix equivalence check. We have presented the theoretical foundation for equivalence checking of quantum circuits in [8] to ensure the correctness of our approach. Quantum circuits often involve *constant inputs*, which refer to inputs with values that are initially fixed to a concrete state, such as $|0\rangle$, at the beginning of the computation. In this paper, we extend the theoretical foundation in [8] to handle the constant inputs for checking the equivalence of quantum circuits. Because the state-of-the-art tools in [3,17] ignore this information, they may lead to incorrect decisions, which will be discussed in Sect. 6. Therefore, our approach enhances the reliability in determining the equivalence problem by taking into account the constant inputs for quantum circuits. We use Maude [4], a high-level specification and programming language based on rewriting logic [13], to develop a support tool called $|QCEC\rangle$ for our approach. Several case studies have been conducted with the tool, and some quantum circuits used to represent Superdense coding as a state transfer [9] have been confirmed to be equivalent if we take into account constant inputs; otherwise, they are not equivalent anymore. These demonstrate the effectiveness of our approach and $|QCEC\rangle$ for the equivalence checking of quantum circuits. $|QCEC\rangle$ and case studies are publicly available at https://github.com/canhminhdo/ket-qcec.

The rest of the paper is organized as follows: Sect. 2 provides basic quantum mechanics and symbolic reasoning for quantum computation based on Dirac notation, Sect. 3 describes the theoretical foundation of equivalence checking of quantum circuits in this work, along with an algorithm constructed based on it, Sect. 4 describes how to specify quantum states, quantum gates, and quantum circuits, Sect. 5 presents how to develop $|QCEC\rangle$ in Maude, Sect. 6 demonstrates how to use $|QCEC\rangle$ to conduct equivalence checking of quantum circuits for some case studies, Sect. 7 presents some existing work, and Sect. 8 concludes the paper with some pieces of future work.

2 Preliminaries

This section briefly describes some basic notations from quantum mechanics based on linear algebra (refer to [15] for more details). Besides, we describe symbolic reasoning [7,22] to reason about quantum computation based on Dirac notation.

2.1 Basic Quantum Mechanics

In classical computing, the fundamental unit of information is a bit whose value is either 0 or 1. In quantum computing, the counterpart is a *quantum bit* or *qubit*, which has two basis states, conventionally written in Dirac notation [6] as $|0\rangle$ and $|1\rangle$, which denote two column vectors $\begin{pmatrix}1\\0\end{pmatrix}$ and $\begin{pmatrix}0\\1\end{pmatrix}$, respectively. In quantum theory, a general state of a quantum system is a superposition or linear combination of basis states. A quantum state is a unit vector in a Hilbert space $\mathcal{H}$, which is a vector space equipped with an inner product such that each Cauchy sequence has a limit. The state of a single qubit is $|\psi\rangle = \alpha|0\rangle + \beta|1\rangle$, where α and β are complex numbers such that $|\alpha|^2 + |\beta|^2 = 1$. States can be represented by column complex vectors as follows: $|\psi\rangle = \begin{pmatrix} \alpha \\ \beta \end{pmatrix} = \alpha|0\rangle + \beta|1\rangle$, where $\{|0\rangle, |1\rangle\}$ forms an orthonormal basis of the two-dimensional complex vector space.

The basis $\{|0\rangle, |1\rangle\}$ is called the *computational* (or *standard*) basis. Besides, there are some other orthonormal bases studied in the literature, such as *diagonal* (or *dual*, or *Hadamard*) basis consisting of the following vectors:

$$|+\rangle = \frac{1}{\sqrt{2}}(|0\rangle + |1\rangle) \text{ and } |-\rangle = \frac{1}{\sqrt{2}}(|0\rangle - |1\rangle)$$

The evolution of a closed quantum system can be performed by a unitary transformation. If the state of a qubit is represented by a column vector, then a unitary transformation U can be represented by a complex-value matrix such that $UU^\dagger = U^\dagger U = I$ or $U^\dagger = U^{-1}$, where $U^\dagger$ is the conjugate transpose of U. U acts on the Hilbert space $\mathcal{H}$ transforming a state $|\psi\rangle$ to a state $|\psi'\rangle$ by a matrix multiplication such that $|\psi'\rangle = U|\psi\rangle$. There are some frequently used quantum gates in applications: the Hadamard gate H, the identity gate I, the Pauli gates X, Y, and Z, and the controlled-NOT gate CX. Note that the CX gate performs on two qubits, while the remaining gates perform on a single qubit. Their matrix representations are as follows:

$$I_2 = \begin{pmatrix} 1 & 0 \\ 0 & 1 \end{pmatrix}, \quad X = \begin{pmatrix} 0 & 1 \\ 1 & 0 \end{pmatrix}, \quad Y = \begin{pmatrix} 0 & -i \\ i & 0 \end{pmatrix},$$

$$Z = \begin{pmatrix} 1 & 0 \\ 0 & -1 \end{pmatrix}, \quad H = \frac{1}{\sqrt{2}}\begin{pmatrix} 1 & 1 \\ 1 & -1 \end{pmatrix}, \quad CX = \begin{pmatrix} 1 & 0 & 0 & 0 \\ 0 & 1 & 0 & 0 \\ 0 & 0 & 0 & 1 \\ 0 & 0 & 1 & 0 \end{pmatrix}.$$

where i is the imaginary unit. For example, the Hadamard gate on a single qubit performs the mapping $|0\rangle \mapsto \frac{1}{\sqrt{2}}(|0\rangle + |1\rangle)$ and $|1\rangle \mapsto \frac{1}{\sqrt{2}}(|0\rangle - |1\rangle)$. The controlled-NOT gate on pairs of qubits, which we explain in the next paragraph, performs the mapping $|00\rangle \mapsto |00\rangle, |01\rangle \mapsto |01\rangle, |10\rangle \mapsto |11\rangle, |11\rangle \mapsto |10\rangle$, which can be understood as inverting the second qubit (referred to as the *target*) if and only if the first qubit (referred to as the *control*) is one.

For multiple qubits, we use the tensor product of Hilbert spaces. Let $\mathcal{H}_1$ and $\mathcal{H}_2$ be two Hilbert spaces. Their tensor product $\mathcal{H}_1 \otimes \mathcal{H}_2$ is defined as a vector space consisting of linear combinations of the vectors $|\psi_1\psi_2\rangle = |\psi_1\rangle|\psi_2\rangle = |\psi_1\rangle \otimes |\psi_2\rangle$, where $|\psi_1\rangle \in \mathcal{H}_1$ and $|\psi_2\rangle \in \mathcal{H}_2$. Systems of two or more qubits may be in *entangled* states, meaning that states of qubits are correlated and inseparable. Entanglement shows that an entangled state of two qubits cannot be expressed as a tensor product of single-qubit states. We can use H and CX gates to create entangled states as follows: $CX((H \otimes I)|00\rangle) = \frac{1}{\sqrt{2}}(|00\rangle + |11\rangle)$.

Let $|\psi\rangle$ be a quantum state and $\theta \in [0, 2\pi)$. In quantum mechanics, the state $e^{i\theta}|\psi\rangle$ is considered to be physically equal to $|\psi\rangle$ with respect to the global phase factor $e^{i\theta}$. Additionally, $\langle\psi|$ is the dual of $|\psi\rangle$ such that $\langle\psi|^\dagger = |\psi\rangle$ and $|\psi\rangle^\dagger = \langle\psi|$. From an observable perspective, two states are undistinguishable if they differ only by a global phase. We can use density matrices $|\psi\rangle\langle\psi|$ to present quantum states $|\psi\rangle$ from which we can eliminate the global phase factor as follows: $e^{i\theta}|\psi\rangle (e^{i\theta}|\psi\rangle)^\dagger = e^{i\theta}|\psi\rangle e^{-i\theta}\langle\psi| = |\psi\rangle\langle\psi|$.

2.2 Symbolic Reasoning

We proposed symbolic reasoning [7] based on Dirac notation with scalars, used a set of laws from quantum mechanics and basic matrix operations to reason about quantum computation in Maude. This section briefly describes terms used in our symbolic reasoning and a set of laws used to reduce terms.

Terms Terms are built from scalars and basis vectors with some operations.

- Scalars are complex numbers. We extend rational numbers supported in Maude to deal with complex numbers. Some operations for scalars, such as multiplication, fraction, addition, conjugation, absolute, power, and square root are specified, but we do not mention them here to make the paper concise.
- Basis vectors are the computational basis written in Dirac notation as $|\mathbf{0}\rangle$ and $|\mathbf{1}\rangle$.
- Operations for matrices consist of scalar multiplication of matrices $\cdot$, matrix product $\times$, matrix addition $+$, tensor product $\otimes$, and the conjugate transpose $\boldsymbol{A}^\dagger$ of a matrix $\boldsymbol{A}$.

In Dirac notation, $\langle\mathbf{0}|$ is the dual of $|\mathbf{0}\rangle$ such that $\langle\mathbf{0}|^\dagger = |\mathbf{0}\rangle$ and $|\mathbf{0}\rangle^\dagger = \langle\mathbf{0}|$; similarly for $\langle\mathbf{1}|$. The terms $|\mathbf{j}\rangle \times \langle\mathbf{k}|$ and $\langle\mathbf{j}| \times |\mathbf{k}\rangle$ may be written shortly as $|\mathbf{j}\rangle\langle\mathbf{k}|$ and $\langle\mathbf{j}|\mathbf{k}\rangle$ for any $\mathbf{j},\mathbf{k} \in \{\mathbf{0},\mathbf{1}\}$. Note that we deal with the inner product $\langle\mathbf{j}|\mathbf{k}\rangle$ by means of $\langle\mathbf{j}| \times |\mathbf{k}\rangle$ in our specification because its result is either zero or one scalar and so its specification is either zero matrix $\boldsymbol{O}$ or identity matrix $\boldsymbol{I}$ (see some laws applied for them in Table 1), respectively. By using these notations, we can intuitively explain how quantum operations work. For example, the $\boldsymbol{X}$ gate performs mapping $|\mathbf{0}\rangle \mapsto |\mathbf{1}\rangle$ and $|\mathbf{1}\rangle \mapsto |\mathbf{0}\rangle$. Therefore, we specify the $\boldsymbol{X}$ gate as $|\mathbf{0}\rangle\langle\mathbf{1}| + |\mathbf{1}\rangle\langle\mathbf{0}|$ in Maude instead of using explicitly the matrix representation $\begin{pmatrix} 0 & 1 \\ 1 & 0 \end{pmatrix}$. Using Dirac notation instead of explicitly complex vectors and matrices as Paykin et al. proposed in [16], making our representations more compact. We have $\boldsymbol{X}|\mathbf{0}\rangle = |\mathbf{1}\rangle\langle\mathbf{0}|\mathbf{0}\rangle + |\mathbf{0}\rangle\langle\mathbf{1}|\mathbf{0}\rangle = |\mathbf{1}\rangle$ because of the use of some laws in Table 1 and similarly for $\boldsymbol{X}|\mathbf{1}\rangle$.

We conventionally specify some basic matrices $\boldsymbol{B}_i$ for $i \in [0..3]$ as follows:

$$\boldsymbol{B}_0 = |\mathbf{0}\rangle \times \langle\mathbf{0}|,\ \boldsymbol{B}_1 = |\mathbf{0}\rangle \times \langle\mathbf{1}|,\ \boldsymbol{B}_2 = |\mathbf{1}\rangle \times \langle\mathbf{0}|,\ \boldsymbol{B}_3 = |\mathbf{1}\rangle \times \langle\mathbf{1}|.$$

The $\boldsymbol{I}, \boldsymbol{X}, \boldsymbol{Y}, \boldsymbol{Z}, \boldsymbol{H}$, and $\boldsymbol{CX}$ gates are then a linear combination of the matrices $\boldsymbol{B}_i$ with scalars as follows:

$I = B_0 + B_3, \quad X = B_1 + B_2, \quad Y = (-i) \cdot B_1 + i \cdot B_2, \quad Z = B_1 + (-1) \cdot B_3,$
$H = \frac{1}{\sqrt{2}} \cdot B_0 + \frac{1}{\sqrt{2}} \cdot B_1 + \frac{1}{\sqrt{2}} \cdot B_2 + (-\frac{1}{\sqrt{2}}) \cdot B_3, \quad CX = B_0 \otimes I_2 + B_3 \otimes X.$

For any $2^n \times 2^n$ matrix, we can represent it as the linear combination of the tensor products of basic matrices B_i with scalars with respect to the number n of qubits. Therefore, quantum states and quantum gates can be represented as terms with Dirac notation and scalars.

Table 1. A set of laws used for symbolic reasoning

No.	Law
L1	$\langle 0\|0 \rangle = \langle 1\|1 \rangle = 1, \langle 1\|0 \rangle = \langle 0\|1 \rangle = 0$
L2	Associativity of $\times, +, \otimes$ and Commutativity of $+$
L3	$0 \cdot A_{m \times n} = O_{m \times n}, \; c \cdot O = O, \; 1 \cdot A = A$
L4	$c \cdot (A + B) = c \cdot A + c \cdot B$
L5	$c_1 \cdot A + c_2 \cdot A = (c_1 + c_2) \cdot A$
L6	$c_1 \cdot (c_2 \cdot A) = (c_1 \cdot c_2) \cdot A$
L7	$(c_1 \cdot A) \times (c_2 \cdot B) = (c_1 \cdot c_2) \cdot (A \times B)$
L8	$A \times (c \cdot B) = (c \cdot A) \times B = c \cdot (A \times B)$
L9	$A \otimes (c \cdot B) = (c \cdot A) \otimes B = c \cdot (A \otimes B)$
L10	$O_{m \times n} \times A_{n \times p} = A_{m \times n} \times O_{n \times p} = O_{m \times p}$
L11	$I_m \times A_{m \times n} = A_{m \times n} \times I_n = A_{m \times n}$
L12	$A + O = O + A = O$
L13	$O_{m \times n} \otimes A_{p \times q} = A_{p \times q} \otimes O_{m \times n} = O_{mp \times nq}$
L14	$A \times (B + C) = A \times B + A \times C$
L15	$(A + B) \times C = A \times C + B \times C$
L16	$(A \otimes B) \times (C \otimes D) = (A \times C) \otimes (B \times D)$
L17	$A \otimes (B + C) = A \otimes B + A \otimes C$
L18	$(A + B) \otimes C = A \otimes C + B \otimes C$
L19	$(c \cdot A)^\dagger = c^* \cdot A^\dagger, \; (A \times B)^\dagger = B^\dagger \times A^\dagger$
L20	$(A + B)^\dagger = A^\dagger + B^\dagger, \; (A \otimes B)^\dagger = A^\dagger \otimes B^\dagger$
L21	$I_m^\dagger = I_m, O_{m \times n}^\dagger = O_{n \times m}, (A^\dagger)^\dagger = A$
L22	$\|0\rangle^\dagger = \langle 0\|, \; \langle 0\|^\dagger = \|0\rangle, \; \|1\rangle^\dagger = \langle 1\|, \; \langle 1\|^\dagger = \|1\rangle$

Laws

Table 1 shows a set of laws derived from the properties of quantum mechanics and basic matrix operations. The reader interested in their proofs in Coq is referred to [19]. Because $|0\rangle$ and $|1\rangle$ can be viewed as 2×1 matrices, then the laws actually describe matrix calculations with Dirac notation, zero and identity matrices, and scalars. These laws are described by equations in Maude and are used to automatically reduce terms until no more matrix operation is applicable. Some laws dedicated to simplifying the expressions about complex numbers are also specified in Maude by means of equations, but we do not mention them here to make the paper concise.

For example, we would like to reduce the term $\boldsymbol{CX} \times ((\boldsymbol{H} \otimes \boldsymbol{I}) \times |0\rangle \otimes |0\rangle)$ to check whether its result is $\frac{1}{\sqrt{2}} \cdot |0\rangle \otimes |0\rangle + \frac{1}{\sqrt{2}} \cdot |1\rangle \otimes |1\rangle$. The term says that the $\boldsymbol{H}$ gate acts on the first qubit followed by the $\boldsymbol{CX}$ gate where the control and target bits are the first and second qubits, respectively. The reader who is interested in the simplification of the term is referred to [7] for more details.

Using the laws, the term is reduced to a canonical form that is a linear combination of the tensor product of the computational basis with scalars. The whole process is conducted automatically in Maude and the result is the same as expected. The key idea is to reduce the matrix multiplication in the form of $\langle i|j \rangle$ into a scalar and simplify the matrix representation by absorbing ones and eliminating zeros (see the law with label L3). In this manner, our symbolic reasoning about matrices can be conducted by rewriting in Maude instead of explicitly calculating matrices.

3 Equivalence Checking of Quantum Circuits

This section describes the theoretical foundation of equivalence checking of quantum circuits based on our previous work [8] together with an algorithm constructed to handle constant inputs for quantum circuits.

3.1 Theoretical Foundation

We propose a method for checking the equivalence of quantum circuits constructed from quantum gates based on their functionality. We suppose that quantum circuits operate on quantum states in a Hilbert space $\mathcal{H}$ with n qubits. The unitary evolution of quantum systems is described by unitary matrices whose size is $2^n \times 2^n$. We first define the equivalence checking problem of quantum circuits.

Definition 1. (*Equivalence checking problem*) Given two quantum circuits represented by unitary matrices, $U = U_m \ldots U_0$ and $U' = U'_{m'} \ldots U'_0$, the equivalence checking problem of U and U' is asked to check whether $U = e^{i\theta} U'$ for some $\theta \in [0, 2\pi)$.

We call $e^{i\theta}$ a global phase that is physically unobservable [15]. In quantum mechanics, quantum states that differ only by a global phase are physically indistinguishable and equivalent under observation [15].

Hence, we define observable equivalence for quantum states as follows:

Definition 2. (*Observable equivalence for quantum states*)
$|\psi\rangle \approx |\psi'\rangle$ is defined as $|\psi\rangle = e^{i\theta} |\psi'\rangle$ for some $\theta \in [0, 2\pi)$. We may write $|\psi\rangle \approx_\theta |\psi'\rangle$ to make it clear from the context.

To check $|\psi\rangle \approx |\psi'\rangle$, we can check the equality of their density matrices $|\psi\rangle\langle\psi|$ and $|\psi'\rangle\langle\psi'|$ because using density matrices to represent quantum states can eliminate the global phase. It is a handy trick to check the observable equivalence of two quantum states by comparing their density matrices. This result is derived from the following lemma.

Lemma 1. $|\psi\rangle \approx |\psi'\rangle$ if and only if $|\psi\rangle\langle\psi| = |\psi'\rangle\langle\psi'|$.

Proof. See the proof in our previous work [8] (specified in Lemma 3.1). □

Recall to check the equivalence of quantum circuits U and U', we need to check whether $U = e^{i\theta}U'$ for some $\theta \in [0, 2\pi)$. We can use the following lemma to solve this problem.

Lemma 2. *Let U and U' be $2^n \times 2^n$ unitary matrices, then $U = e^{i\theta}U'$ for some $\theta \in [0, 2\pi)$ if and only if $U|\psi\rangle \approx_\theta U'|\psi\rangle$ for any vector $|\psi\rangle \in \mathcal{H}$.*

Proof. See the proof in our previous work [8] (specified in Lemma 3.2). □

In Lemma 2, it is unfeasible to consider any vector $|\psi\rangle \in \mathcal{H}$ because there are infinite vectors in $\mathcal{H}$. Therefore, we introduce the following lemma to help us check whether $U = e^{i\theta}U'$ by considering only basis vectors in an orthonormal basis of $\mathcal{H}$. If the dimension of $\mathcal{H}$ is n, we need to consider at most 2^n basis vectors with respect to the same global phase.

Lemma 3. *Let U and U' be $2^n \times 2^n$ matrices, then $U = e^{i\theta}U'$ for some $\theta \in [0, 2\pi)$ if and only if $U|\phi_i\rangle \approx_\theta U'|\phi_i\rangle$ for each basis vector $|\phi_i\rangle$ in an orthonormal basis of $\mathcal{H}$.*

Proof. See the proof in our previous work [8] (specified in Lemma 3.3). □

Remark 1. Checking $U|\psi_i\rangle \approx_\theta U'|\psi_i\rangle$ is actually checking the observable equivalence of the ith column vector of U and the ith column vector of U' with respect to the phase θ. In order to do so, we have the following lemma.

Lemma 4. *$U|\phi_i\rangle \approx_\theta U'|\phi_i\rangle$ for each basis vector $|\phi_i\rangle$ in an orthonormal basis of $\mathcal{H}$ if and only if $U|\phi_i\rangle \approx U'|\phi_i\rangle$ for each $|\phi_i\rangle$ and $U|\phi_i\rangle(U|\phi_j\rangle)^\dagger = U'|\phi_i\rangle(U'|\phi_j\rangle)^\dagger$ for each $|\phi_i\rangle$ and $|\phi_j\rangle$.*

Proof. See the proof in our previous work [8] (specified in Lemma 3.4). □

Now, we introduce our main theorem to check whether $U = e^{i\theta}U'$ for some $\theta \in [0, 2\pi)$ as follows.

Theorem 1. *Let U and U' be $2^n \times 2^n$ matrices, then $U = e^{i\theta}U'$ for some $\theta \in [0, 2\pi)$ if and only if $U|\phi_i\rangle(U|\phi_i\rangle)^\dagger = U'|\phi_i\rangle(U'|\phi_i\rangle)^\dagger$ for each basis vector $|\phi_i\rangle$ and $U|\phi_i\rangle(U|\phi_j\rangle)^\dagger = U'|\phi_i\rangle(U'|\phi_j\rangle)^\dagger$ for each $|\phi_i\rangle$ and $|\phi_j\rangle$ in an orthonormal basis of $\mathcal{H}$.*

Proof. See the proof in our previous work [8] (specified in Theorem 3.5). □

It is extremely expensive to calculate matrix-matrix multiplications $U_m \ldots U_0$ and $U'_{m'} \ldots U'_0$ to obtain U and U' and subsequently multiple with each $|\phi_i\rangle$ and $|\phi_j\rangle$ in Theorem 1 because of the exponential size of unitary matrices. Instead, we can perform a series of matrix-vector multiplications between unitary matrices and vectors in sequence as follows:

$$|u_i^0\rangle = U_0 |\phi_i\rangle, |u_i^1\rangle = U_1 |u_i^0\rangle, \ldots, |u_i^m\rangle = U_m \cdot u_i^{m-1}$$

The ith column vector of matrix U is $|u_i\rangle$ (i.e., $|u_i^m\rangle$) and similarly for the ith column vector $|u'_i\rangle$ of matrix U'. We are now ready to check whether $|u_i\rangle\langle u_i|$ is equal to $|u'_i\rangle\langle u'_i|$ for the first condition in Theorem 1. Moreover, for the second condition in Theorem 1, it suffices to fix $|u_i\rangle$ and $|u'_i\rangle$, and check whether $|u_i\rangle\langle u_i| u_j = |u'_i\rangle\langle u'_i| u'_j$ for all $j \neq i$. This is an efficient way to handle the calculation in Theorem 1.

3.2 Handling Constant Inputs for Equivalence Checking

Quantum circuits often involve constant inputs, which refer to inputs with values that are initially fixed to a concrete state, such as $|0\rangle$, at the beginning of the computation. Some quantum circuits are equivalent only if we consider the constant inputs, as presented in Sect. 6. This section describes how to handle constant inputs for the equivalence checking of quantum circuits.

Let n be the number of qubits among which m ($\leq n$) qubits are initially fixed to the state $|0\rangle$. We refer to these values as the constant inputs for m qubits. To check the equivalence of quantum circuits, we only need to consider 2^{n-m} basis vectors in the orthonormal basis of $\mathcal{H}$ with n dimension. Let $A = \{|\phi_0\rangle, \ldots, |\phi_{2^n-1}\rangle\}$ be an orthonormal basis, where $|\phi_i\rangle$ is in the form of $|b_0 \ldots b_{n-1}\rangle$ with $b_j \in \{0,1\}$ for $i \in [0 \ldots 2^n - 1], j \in [0 \ldots n-1]$. Let $idx(|\phi_i\rangle, j)$ be the value of b_j in basis vector $|\phi_i\rangle$. Let $const$ be the set of indices denoting the places where constant inputs are used for quantum circuits. The subset consisting of 2^{n-m} basis vectors that we need to consider for checking the equivalence problem is defined as follows:

$$B = \{|\phi_i\rangle \in A \mid \forall j \in const, idx(|\phi_i\rangle, j) \neq 1\}$$

We have the following lemma to support our claim.

Lemma 5. *Let U and U' be $2^n \times 2^n$ matrices and const be the set of indices for constant inputs, then $U = e^{i\theta} U'$ for some $\theta \in [0, 2\pi)$ if and only if $U |\phi_i\rangle \approx_\theta U' |\phi_i\rangle$ for each basis vector $|\phi_i\rangle$ in B.*

Proof. The '*only if*' part (the $\Rightarrow$ direction) is immediate from Lemma 2.

For the '*if*' part (the $\Leftarrow$ direction), we have the following. For arbitrary $|\psi\rangle \in \mathcal{H}$, there exists $c_0, \ldots, c_i, \ldots, c_{2^n-1}$ such that $|\psi\rangle = c_0 |\phi_0\rangle + \cdots + c_i |\phi_i\rangle + \cdots + c_{2^n-1} |\phi_{2^n-1}\rangle$ and $|c_0|^2 + \cdots + |c_i|^2 + \cdots + |c_{2^n-1}|^2 = 1$. Because of the constant inputs used for $|\psi\rangle$, the qubits of $|\psi\rangle$ at indices belonging to $const$ are always $|0\rangle$. Therefore, for each $|\phi_i\rangle \in A \setminus B$, its coefficient c_i above has to be 0.

Algorithm 1: Equivalence Checking of Quantum Circuits

input : n – the dimension of a Hilbert space
$U = U_m \ldots U_0$ and $U' = U'_{m'} \ldots U'_0$ – two quantum circuits
$\{|\phi_0\rangle, \ldots, |\phi_{2^n-1}\rangle\}$ – an orthonormal basis of a Hilbert space $\mathcal{H}$
$\theta \in [0, 2\pi)$ – the phase
$const$ – constant inputs
output: True ($U = e^{i\theta}U'$) or False ($U \neq e^{i\theta}U'$)

1 **forall** the $|\phi_i\rangle \in \{|\phi_0\rangle, \ldots, |\phi_{2^n-1}\rangle\}$ **do**
2   **if** $\neg isNecessary(|\phi_i\rangle, const)$ **then**
3 | continue
4     $|u_i\rangle = U_m \cdot \ldots \cdot U_0 \cdot |\phi_i\rangle$
5     $|u'_i\rangle = U'_{m'} \cdot \ldots \cdot U'_0 \cdot |\phi_i\rangle$
6   **if** $|u_i\rangle\langle u_i| \neq |u'_i\rangle\langle u'_i|$ **then**
7 | **return** False
8   **if** $i \neq 0 \wedge |u_0\rangle\langle u_0| u_i \neq |u'_0\rangle\langle u'_0| u'_i$ **then**
9 | **return** False
10 **return** True

For each basis vector $\phi_i \in B$, we have $U|\phi_i\rangle \approx_\theta U'|\phi_i\rangle$ from the assumption. From Definition 2, we have $U|\phi_i\rangle = e^{i\theta}U'|\phi_i\rangle$ for some $\theta \in [0, 2\pi)$. Therefore, for any complex number c_i, we have $Uc_i|\phi_i\rangle = e^{i\theta}U'c_i|\phi_i\rangle$ for each $|\phi_i\rangle \in B$. Furthermore, for each basis vector $\phi_i \in A \setminus B$, we also have $Uc_i|\phi_i\rangle = e^{i\theta}U'c_i|\phi_i\rangle$ for any $\theta \in [0, 2\pi)$ because c_i is 0.

To this end, we have the following:

$$\begin{aligned}
U|\psi\rangle &= Uc_0|\phi_0\rangle + \cdots + Uc_i|\phi_i\rangle + \cdots + Uc_{2^n-1}|\phi_{2^n-1}\rangle \\
&= e^{i\theta}U'c_0|\phi_0\rangle + \cdots + e^{i\theta}U'c_i|\phi_i\rangle + \cdots + e^{i\theta}U'c_{2^n-1}|\phi_{2^n-1}\rangle \\
&= e^{i\theta}U'(c_0|\phi_0\rangle + \cdots + c_i|\phi_i\rangle + \cdots + c_{2^n-1}|\phi_{2^n-1}\rangle) \\
&= e^{i\theta}U'|\psi\rangle
\end{aligned}$$

for any $|\psi\rangle$ and θ. It means that $U|\psi\rangle \approx_\theta U'|\psi\rangle$ for any vector $|\psi\rangle$. Therefore, $U = e^{i\theta}U'$ by Lemma 2. □

3.3 An Algorithm for Equivalence Checking of Quantum Circuits

An algorithm for equivalence checking of quantum circuits can be constructed based on Theorem 1 and Lemma 5, which is shown as Algorithm 1. Given two quantum circuits in the form of $U = U_m \ldots U_0$ and $U' = U'_{m'} \ldots U'_0$, an orthonormal basis $\{|\phi_0\rangle, \ldots, |\phi_{2^n-1}\rangle\}$, and constant inputs, for each basis vector ϕ_i, we first check whether it is necessary to consider this basis vector or not based on Lemma 5 in the code fragment at lines 2–3. If not, we skip this basis vector and continue with the next one. Otherwise, we construct the series of matrix-vector multiplications between unitary matrices and vectors with right associativity to obtain $|u_i\rangle$ and $|u'_i\rangle$ in the code fragment at lines 4–5. We then check whether their corresponding density matrices $|u_i\rangle\langle u_i|$ and $|u'_i\rangle\langle u'_i|$ are equal for the first condition in Theorem 1 in the code fragment at lines 6–7. If this is not the case, False is returned. Otherwise, we keep on checking for the second condition in Theorem 1 except for the case of the basis vector $|\phi_0\rangle$ in the code fragment at lines 8–9. If $|u_0\rangle\langle u_0|\, u_i$ is not equal to $|u'_0\rangle\langle u'_0|\, u'_i$, False is returned. Otherwise, we move to check for other basis vectors. True is returned at the end once all basis vectors have been checked.

4 Formal Specification

This section shows how we specify qubits, quantum gates, and quantum circuits in Maude.

4.1 Specification of Qubits and Quantum Gates

Qubits are specified as the linear combination of tensor product of the computational basis in Dirac notation with scalars and similarly for quantum gates. Because $|0\rangle$ and $|1\rangle$ can be viewed as 2×1 matrices, then qubits and quantum gates are basically matrices. Quantum gates act on qubits (a quantum state) specified as a matrix multiplication.

4.2 Specification of Quantum Circuits

Quantum circuits without measurements are composed of a sequence of quantum gates. We specify it as a list of actions in which each action is one of the forms as follows:

- I(i) applies the ***I*** gate on q_i,
- X(i) applies the ***X*** gate on q_i,
- Y(i) applies the ***Y*** gate on q_i,
- Z(i) applies the ***Z*** gate on q_i,
- H(i) applies the ***H*** gate on q_i,
- S(i) applies the ***S*** gate on q_i,
- T(i) applies the ***T*** gate on q_i,
- CX(i,j) applies the ***CX*** gate on q_i and q_j,
- CY(i,j) applies the ***CY*** gate on q_i and q_j,
- CZ(i,j) applies the ***CZ*** gate on q_i and q_j,
- SWAP(i,j) applies the ***SWAP*** gate on q_i and q_j,
- CCX(i,j,k) applies the ***CCX*** gate on q_i, q_j, and q_k,
- CCZ(i,j,k) applies the ***CCZ*** gate on q_i, q_j, and q_k,
- CSWAP(i,j,k) applies the ***CSWAP*** gate on q_i, q_j, and q_k,

Note that we consider formalizing the most commonly used quantum gates to describe the behavior of quantum circuits. Based on the quantum gates specified above, we can easily describe the behavior of quantum circuits. For example, the second circuit of Superdense coding as a state transfer in Fig. 1 in Sect. 6 is expressed as a list of actions: CX(1,3) H(2) CZ(0,2) H(2).

Let Π be a fixed set of quantum gates (or actions) specified above and $\mathcal{U}(\mathbb{C}^{2^n})$ be a set of $2^n \times 2^n$ unitary matrices, where n is the number of qubits. A function $\bar{v} : \Pi \to \mathcal{U}(\mathbb{C}^{2^n})$ is a mapping from actions to unitary matrices as follows:

$$\bar{v}(\mathtt{I}(i)) = \boldsymbol{I}^{\otimes i} \otimes \boldsymbol{I} \otimes \boldsymbol{I}^{\otimes n-i-1},$$

$$\bar{v}(\mathtt{H}(i)) = \boldsymbol{I}^{\otimes i} \otimes \boldsymbol{H} \otimes \boldsymbol{I}^{\otimes n-i-1}, \quad \bar{v}(\mathtt{X}(i)) = \boldsymbol{I}^{\otimes i} \otimes \boldsymbol{X} \otimes \boldsymbol{I}^{\otimes n-i-1},$$

$$\bar{v}(\mathtt{Y}(i)) = \boldsymbol{I}^{\otimes i} \otimes \boldsymbol{Y} \otimes \boldsymbol{I}^{\otimes n-i-1}, \quad \bar{v}(\mathtt{Z}(i)) = \boldsymbol{I}^{\otimes i} \otimes \boldsymbol{Z} \otimes \boldsymbol{I}^{\otimes n-i-1},$$

$$\bar{v}(\mathtt{S}(i)) = \boldsymbol{I}^{\otimes i} \otimes |0\rangle\langle 0| \otimes \boldsymbol{I}^{\otimes n-i-1} + \boldsymbol{I}^{\otimes i} \otimes i \cdot |1\rangle\langle 1| \otimes \boldsymbol{I}^{\otimes n-i-1},$$

$$\bar{v}(\mathtt{T}(i)) = \boldsymbol{I}^{\otimes i} \otimes |0\rangle\langle 0| \otimes \boldsymbol{I}^{\otimes n-i-1} + \boldsymbol{I}^{\otimes i} \otimes (1+i)/\sqrt{2} \cdot |1\rangle\langle 1| \otimes \boldsymbol{I}^{\otimes n-i-1},$$

$$\bar{v}(\mathtt{CX}(i,j)) = \boldsymbol{I}^{\otimes i} \otimes |0\rangle\langle 0| \otimes \boldsymbol{I}^{\otimes n-i-1} + (\boldsymbol{I}^{\otimes i} \otimes |1\rangle\langle 1| \otimes \boldsymbol{I}^{\otimes n-i-1})(\boldsymbol{I}^{\otimes j} \otimes \boldsymbol{X} \otimes \boldsymbol{I}^{\otimes n-j-1}),$$

$$\bar{v}(\mathtt{CY}(i,j)) = \boldsymbol{I}^{\otimes i} \otimes |0\rangle\langle 0| \otimes \boldsymbol{I}^{\otimes n-i-1} + (\boldsymbol{I}^{\otimes i} \otimes |1\rangle\langle 1| \otimes \boldsymbol{I}^{\otimes n-i-1})(\boldsymbol{I}^{\otimes j} \otimes \boldsymbol{Y} \otimes \boldsymbol{I}^{\otimes n-j-1}),$$

$$\bar{v}(\mathtt{CZ}(i,j)) = \boldsymbol{I}^{\otimes i} \otimes |0\rangle\langle 0| \otimes \boldsymbol{I}^{\otimes n-i-1} + (\boldsymbol{I}^{\otimes i} \otimes |1\rangle\langle 1| \otimes \boldsymbol{I}^{\otimes n-i-1})(\boldsymbol{I}^{\otimes j} \otimes \boldsymbol{Z} \otimes \boldsymbol{I}^{\otimes n-j-1}),$$

$$\bar{v}(\mathtt{SWAP}(i,j)) = \bar{v}(\mathtt{CX}(i,j)) \times \bar{v}(\mathtt{CX}(j,i)) \times \bar{v}(\mathtt{CX}(i,j)),$$

$$\bar{v}(\mathtt{CCX}(i,j,k)) = \boldsymbol{I}^{\otimes i} \otimes |0\rangle\langle 0| \otimes \boldsymbol{I}^{\otimes j-i-1} \otimes \boldsymbol{I} \otimes \boldsymbol{I}^{\otimes k-j-1} \otimes \boldsymbol{I} \otimes \boldsymbol{I}^{\otimes n-k-1}$$
$$+ \boldsymbol{I}^{\otimes i} \otimes |1\rangle\langle 1| \otimes \boldsymbol{I}^{\otimes j-i-1} \otimes |0\rangle\langle 0| \otimes \boldsymbol{I}^{\otimes k-j-1} \otimes \boldsymbol{I} \otimes \boldsymbol{I}^{\otimes n-k-1},$$
$$+ \boldsymbol{I}^{\otimes i} \otimes |1\rangle\langle 1| \otimes \boldsymbol{I}^{\otimes j-i-1} \otimes |1\rangle\langle 1| \otimes \boldsymbol{I}^{\otimes k-j-1} \otimes |0\rangle\langle 0|1 \otimes \boldsymbol{I}^{\otimes n-k-1},$$
$$+ \boldsymbol{I}^{\otimes i} \otimes |1\rangle\langle 1| \otimes \boldsymbol{I}^{\otimes j-i-1} \otimes |1\rangle\langle 1| \otimes \boldsymbol{I}^{\otimes k-j-1} \otimes |1\rangle\langle 1|0 \otimes \boldsymbol{I}^{\otimes n-k-1},$$

$$\bar{v}(\mathtt{CCZ}(i,j,k)) = \boldsymbol{I}^{\otimes i} \otimes |0\rangle\langle 0| \otimes \boldsymbol{I}^{\otimes j-i-1} \otimes \boldsymbol{I} \otimes \boldsymbol{I}^{\otimes k-j-1} \otimes \boldsymbol{I} \otimes \boldsymbol{I}^{\otimes n-k-1}$$
$$+ \boldsymbol{I}^{\otimes i} \otimes |1\rangle\langle 1| \otimes \boldsymbol{I}^{\otimes j-i-1} \otimes |0\rangle\langle 0| \otimes \boldsymbol{I}^{\otimes k-j-1} \otimes \boldsymbol{I} \otimes \boldsymbol{I}^{\otimes n-k-1},$$
$$+ \boldsymbol{I}^{\otimes i} \otimes |1\rangle\langle 1| \otimes \boldsymbol{I}^{\otimes j-i-1} \otimes |1\rangle\langle 1| \otimes \boldsymbol{I}^{\otimes k-j-1} \otimes |0\rangle\langle 0| \otimes \boldsymbol{I}^{\otimes n-k-1},$$
$$- \boldsymbol{I}^{\otimes i} \otimes |1\rangle\langle 1| \otimes \boldsymbol{I}^{\otimes j-i-1} \otimes |1\rangle\langle 1| \otimes \boldsymbol{I}^{\otimes k-j-1} \otimes |1\rangle\langle 1| \otimes \boldsymbol{I}^{\otimes n-k-1},$$

$$\bar{v}(\mathtt{CSWAP}(i,j,k)) = \boldsymbol{I}^{\otimes i} \otimes |0\rangle\langle 0| \otimes \boldsymbol{I}^{\otimes j-i-1} \otimes \boldsymbol{I} \otimes \boldsymbol{I}^{\otimes k-j-1} \otimes \boldsymbol{I} \otimes \boldsymbol{I}^{\otimes n-k-1}$$
$$+ \boldsymbol{I}^{\otimes i} \otimes |1\rangle\langle 1| \otimes \boldsymbol{I}^{\otimes j-i-1} \otimes |0\rangle\langle 0| \otimes \boldsymbol{I}^{\otimes k-j-1} \otimes |0\rangle\langle 0| \otimes \boldsymbol{I}^{\otimes n-k-1},$$
$$+ \boldsymbol{I}^{\otimes i} \otimes |1\rangle\langle 1| \otimes \boldsymbol{I}^{\otimes j-i-1} \otimes |0\rangle\langle 0|1 \otimes \boldsymbol{I}^{\otimes k-j-1} \otimes |1\rangle\langle 1|0 \otimes \boldsymbol{I}^{\otimes n-k-1},$$
$$+ \boldsymbol{I}^{\otimes i} \otimes |1\rangle\langle 1| \otimes \boldsymbol{I}^{\otimes j-i-1} \otimes |1\rangle\langle 1|0 \otimes \boldsymbol{I}^{\otimes k-j-1} \otimes |1\rangle\langle 1|0 \otimes \boldsymbol{I}^{\otimes n-k-1},$$

where

$$\boldsymbol{I}^{\otimes i} = \overbrace{\boldsymbol{I} \otimes \cdots \otimes \boldsymbol{I}}^{i}.$$

Note that $|0\rangle\langle 0|, |0\rangle\langle 1|, |1\rangle\langle 0|$ and $|1\rangle\langle 1|$ are represented by $\boldsymbol{B}_0, \boldsymbol{B}_1, \boldsymbol{B}_2$, and $\boldsymbol{B}_3$ from the definition in Sect. 2.2. If we replace $\boldsymbol{I}, \boldsymbol{X}, \boldsymbol{Y}, \boldsymbol{Z}$, and $\boldsymbol{H}$ above by the linear combination of $\boldsymbol{B}_0, \boldsymbol{B}_1, \boldsymbol{B}_2$, and $\boldsymbol{B}_3$ as defined in Sect. 2.2, we can reduce each unitary quantum gate to a canonical form of a linear combination of the tensor product of $\boldsymbol{B}_0, \boldsymbol{B}_1, \boldsymbol{B}_2$, and $\boldsymbol{B}_3$ (i.e., $|0\rangle\langle 0|, |0\rangle\langle 1|, |1\rangle\langle 0|$ and $|1\rangle\langle 1|$) by using the laws in Sect. 2.2.

Theorem 2. *Each unitary matrix is represented by a canonical form of a linear combination of the tensor product of $|0\rangle\langle 0|, |0\rangle\langle 1|, |1\rangle\langle 0|$ and $|1\rangle\langle 1|$ with scalars with respect to the number n of qubits in a Hilbert space $\mathcal{H}$.*

Proof. Each tensor product of $|0\rangle\langle 0|, |0\rangle\langle 1|, |1\rangle\langle 0|$ and $|1\rangle\langle 1|$ with respect to the number n of qubits produces a unique $2^n \times 2^n$ matrix where only one distinct element of the matrix is one and the others are zero. Therefore, any unitary matrix can be represented by a canonical form of a linear combination of the tensor product of $|0\rangle\langle 0|, |0\rangle\langle 1|, |1\rangle\langle 0|$ and $|1\rangle\langle 1|$ with scalars. □

Using Theorem 2, we can use symbolic reasoning in Sect. 2.2 to reduce any unitary matrix or multiplication of unitary matrices to a canonical form, which is a linear combination of the tensor product of $|0\rangle\langle 0|, |0\rangle\langle 1|, |1\rangle\langle 0|$ and $|1\rangle\langle 1|$ with scalars with respect to the number of qubits used in quantum circuits.

Theorem 3. *Each quantum state is represented by a canonical form of a linear combination of the tensor product of computational basis vectors $|0\rangle$ and $|1\rangle$ with scalars with respect to the number n of qubits in a Hilbert space $\mathcal{H}$.*

Proof. It is immediate from the definition of a quantum state. □

Based on Theorems 2 and 3, the multiplication of a quantum gate and a quantum state produces a new quantum state, which also has to be in a canonical form.

5 A Support Tool

We use Maude [4], a high-level specification/programming language based on rewriting logic [13], to develop a support tool called |QCEC⟩ for the equivalence checking of quantum circuits. Symbolic reasoning described in Sect. 2.2 is adopted in |QCEC⟩ for reasoning about quantum computation, which actually is matrix calculation with Dirac notation and scalars under the laws of quantum mechanics and matrix operations. The specification of quantum states, quantum gates, and quantum circuits is specified in |QCEC⟩. The function $\bar{v}$ is also developed to map an action to an actual unitary matrix in Dirac notation.

We are now ready to develop the function equivCheck to determine the equivalence of two quantum circuits based on Algorithm 1. For brevity, we only describe the specification for equivCheck and the Maude syntax is used without detailed explanation as follows:

```
op equivCheck : ActionList ActionList Info -> Bool .
ceq equivCheck(AL, AL', IF) = checkWithBasis(AL, AL', IF, N)
if N := findN(AL AL') .
```

equivCheck takes as inputs two action lists AL and AL' of sort ActionList and possibly constant inputs IF of sort Info, and then returns a Boolean value indicating the result of the equivalence problem. We first find the number N of qubits using the function findN and then begin checking with the function checkWithBasis, which is defined as follows:

op checkWithBasis : ActionList ActionList Info Nat -> Bool .
eq checkWithBasis(AL,AL',IF,N) = firstCheck(AL,AL',IF,N,2^N,0) .

checkWithBasis takes as inputs AL, AL' , IF, and N, and then initializes the number MAX (i.e., 2^N) of basis vectors and considers the first basis vector associated with index 0 for the function firstCheck, which is defined as follows:

op firstCheck : ActionList ActionList Info Nat Nat Nat
                 -> Bool .
ceq firstCheck(AL, AL', IF, N, MAX, 0) =
    if (V0 x (V0)^+) == (V0' x (V0')^+) **then**
        nextCheck(AL, AL', IF, N, MAX, 1, V0, V0')
    else false **fi**
if Phi := basis(0, N) /\ V0 := calWithBasis(AL, Phi)
/\ V0' := calWithBasis(AL', Phi) .

where (V0)^+ and (V0')^+ denote the conjugate transpose of V0 and V0', respectively. firstCheck is the initial checking step for the equivalence problem by considering the first basis vector. We prepare the basis vector Phi (i.e., $|\phi_0\rangle$ in Algorithm 1) using the function basis with the given inputs 0 and N; calculate the vector V0 (i.e., $|u_0\rangle$ in Algorithm 1) using the function calWithBasis with the given inputs AL and Phi; and perform a similar process for the vector V0' (i.e., $|u'_0\rangle$ in Algorithm 1). We then check whether the density matrices of V0 and V0' are equal. If they are not, false is returned. Otherwise, we move to check for the next basis vector associated with index 1 using the function nextCheck. Because we need to consider the same global phase for all basis vectors, V0 and V0' are also passed to nextCheck, which is defined as follows:

op nextCheck : ActionList ActionList Info Nat Nat Nat Vect
                Vect -> Bool .
ceq nextCheck(AL, AL', IF, N, MAX, IDX, V0, V0') =
    if (isNecessary(IDX, N, getConst(IF))) **then**
        basisCheck(AL, AL', IF, N, MAX, IDX, V0, V0')
    else nextCheck(AL, AL', IF, N, MAX, IDX + 1, V0, V0') **fi**
if IDX < MAX .
eq nextCheck(AL,AL',IF,N,MAX,IDX,V0,V0') = true [owise] .

nextCheck examines the equivalence problem by considering the current basis vector associated with the index IDX. If IDX is not less than MAX, meaning that all basis vectors have been checked, true is returned. Otherwise, we check whether the current basis vector should be considered according to the constant inputs using the function isNecessary, where the function getConst is used to obtain the constant input information. If that is not the case, we move to check the next basis vector by incrementing IDX and calling nextCheck again. Otherwise, we check the equivalence problem with the current basis vector using the function basisCheck, which is defined as follows:

```
op basisCheck : ActionList ActionList Info Nat Nat Nat Vect
                Vect -> Bool .
ceq basisCheck(AL, AL', IF, N, MAX, IDX, V0, V0') =
    if (V x (V)^+) == (V' x (V')^+) then
       if (V0 x (V)^+) == (V0' x (V')^+) then
          nextCheck(AL, AL', IF, N, MAX, IDX + 1, V0, V0')
       else false fi
    else false fi
if Phi := basis(IDX, N) /\ V := calWithBasis(AL, Phi)
/\ V'  := calWithBasis(AL', Phi) .
```

basisCheck first calculates the basis vector Phi associated with the index IDX (i.e., $|\phi_i\rangle$ in Algorithm 1), the vector V (i.e., $|u_i\rangle$ in Algorithm 1), and the vector V' (i.e., $|u'_i\rangle$ in Algorithm 1). It then checks whether the density matrices of V and V' are equal, and whether the same global phase is shared between V0 and V, and V0' and V'. If that is the case, we move to check the next basis vector by incrementing IDX and calling nextCheck again. Otherwise, false is returned.

Thereby, the whole process of the function equivCheck for checking the equivalence of two quantum circuits represented by $A_0 \ldots A_m$ and $A'_0 \ldots A'_{m'}$ can be conducted automatically using the following command in Maude:

red equivCheck($A_0 \ldots A_m$, $A'_0 \ldots A'_{m'}$) .

6 Case Studies

We used |QCEC⟩ to confirm the equivalence of many quantum circuits in [9] for various groups of case studies, such as Control Reversal, Distributed CNOT, CNOT Minor, Parallel to Λ CNOT, and Quantum State Transfer. Additionally, we modified their circuits and confirmed their non-equivalence as well. The modification was conducted by randomly dropping a quantum gate or changing the indices to which a quantum gate is applied. The experiments were conducted with an iMac that carries a 4 GHz microprocessor with eight cores and 32 GB memory of RAM. All experiments were completed in a few seconds as shown in Table 2, demonstrating the effectiveness of our approach. |QCEC⟩ and case studies are publicly available at https://github.com/canhminhdo/ket-qcec.

For the sake of simplicity, we only present how to use |QCEC⟩ to confirm the equivalence of some quantum circuits for Superdense coding [2] as a state transfer [9] and also point out in which cases these quantum circuits are not equivalent. The four circuits shown in Fig. 1 were considered equivalent circuits to represent Superdense coding as a state transfer in [9] if the third and fourth qubits are initially fixed to |**0**⟩ at the beginning of the computation. We adopt the specification of quantum circuits presented in Sect. 4 to describe the behavior of quantum circuits as action lists. Therefore, the action lists of the four quantum circuits are described in order as follows:

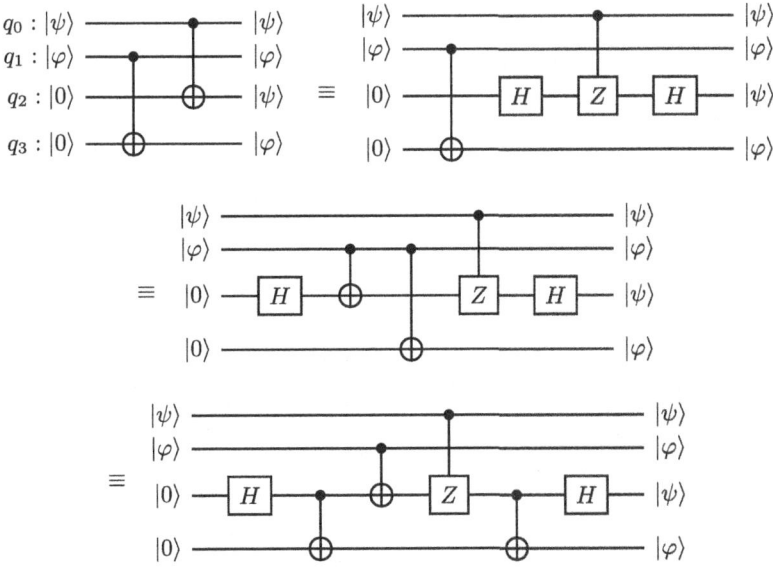

Fig. 1. Superdense coding as a state transfer

```
AL1 = CX(1,3) CX(0,2)
AL2 = CX(1,3) H(2) CZ(0,2) H(2)
AL3 = H(2) CX(1,2) CX(1,3) CZ(0,2) H(2)
AL4 = H(2) CX(2,3) CX(1,2) CZ(0,2) CX(2, 3) H(2)
```

We check whether the four quantum circuits are equivalent by using the following commands:

```
red equivCheck(AL1, AL2) .
red equivCheck(AL3, AL4) .
red equivCheck(AL2, AL3) . --- not equivalent to any inputs
red equivCheck(AL2, AL3, const: (2, 3)) .
```

The first two commands return true and so we can conclude the equivalence of the first and second quantum circuits and the equivalence of the third and fourth quantum circuits with any inputs. The third command returns false, indicating that the second and third quantum circuits are not equivalent with any inputs. However, if we use constant inputs for the third and fourth qubits as reported in [9], the fourth command returns true again. This result confirms the equivalence of the second and third quantum circuits under this condition. The state-of-the-art tools in [3,17] do not take into account the constant inputs, and so they simply conclude that the second and the third quantum circuits are non-equivalent, while they are actually equivalent under the condition that the constant inputs are used. Therefore, using our approach and |QCEC⟩ is more reliable in this situation. These experiments demonstrate the effectiveness of our approach and |QCEC⟩ for checking the equivalence of quantum circuits.

Table 2. Experimental results

Quantum circuits	Qubits	Time for equivalence cases	Time for non-equivalence cases
Control reversal	2	6 ms	1 ms
Distributed CNOT	3	1 ms	≈0 ms
CNOT minor	3	3 ms	1 ms
Parallel to Λ CNOT	3	1 ms	≈0 ms
Quantum state transfer	4	16 ms	4 ms

The experimental results for our case studies are summarized in Table 2. The first and second columns denote the groups of case studies, each of which contains several quantum circuits, and their corresponding number of qubits, respectively. The third column shows the execution time required to confirm the equivalence of quantum circuits used for each group. Similarly, the last column shows the execution time for confirming the non-equivalence of modified quantum circuits, where some gates are randomly dropped or the indices to which a gate is applied are changed. Note that we checked the equivalence of several quantum circuits for each group and selected the longest execution time, as shown in the third and fourth columns. All experiments were completed in a few seconds, demonstrating the effectiveness of our approach and |QCEC⟩ in checking the equivalence of both equivalent and non-equivalent quantum circuits. Although the number of qubits used for quantum circuits in our case studies is small, the execution time for non-equivalence cases is rather smaller than that for equivalence cases. This discrepancy arises because fewer iterations are needed to decide the equivalence problem for the former cases, whereas the latter cases require as many iterations as the number of basis vectors. Therefore, our approach becomes faster than the actual matrix equivalence check, particularly in cases involving non-equivalent quantum circuits. As part of our future work, we aim to extend our symbolic reasoning to handle a broader range of quantum gates, enabling the treatment of more complicated quantum circuits. Additionally, we plan to conduct more case studies involving a larger number of qubits to further demonstrate the effectiveness of our approach and |QCEC⟩.

7 Related Work

There are two main approaches to equivalence checking of quantum circuits: one based on quantum decision diagrams [3] and the other based on the ZX calculus [17].

Burgholzer and Wille [3] have proposed an advanced method for equivalence checking of quantum circuits based on a decision diagram. Their approach involves two quantum circuits $U = U_m \ldots U_0$ and $U'_{m'} \ldots U'_0$ as inputs and they check whether the two quantum circuits are equivalent. They leverage two key observations: (1) quantum circuits are inherently reversible, and (2) even small differences in quantum circuits may impact the overall behavior of quantum circuits. Their strategy is as follows. For (2), they first randomly prepare some basis vectors $|\phi_i\rangle$ and calculate the ith column of each matrix U and U' to obtain $|u_i\rangle$ and $|u_i'\rangle$ as we do. They then compare $|u_i\rangle$ and $|u_i'\rangle$ modulo the global phase by using the fidelity, denoted $\mathcal{F} = |\langle u_i|u_i'\rangle|^2$, to measure the overlap between the two states. The two states are considered equivalent if the fidelity between them is 1 up to a given tolerance ε. This is an approximate estimation, while we use their density matrices for the comparison, which provides an exact estimation. After several runs, if they find $|u_i\rangle$ and $|u_i'\rangle$ that are not equivalent, the process is stopped. Otherwise, they attempt to resolve $U \Rightarrow I \Leftarrow U'$ into the identity matrix I based on (1) to solve the equivalence checking problem. However, calculating $U_i(U_{i'}')^{\dagger}$ involves an expensive matrix-matrix multiplication. We have proven a theorem that it suffices to consider all basis vectors in an orthonormal basis to conclude the equivalence checking problem. Our approach also takes the global phase into account. Additionally, they use a decision diagram to encode quantum states and quantum gates, while we use Dirac notation. Disregarding the encoding is used, our strategy to check the equivalence of quantum circuits can be adopted by other approaches.

Peham et al. [17] proposed equivalence checking of quantum circuits with the ZX calculus. The ZX calculus is a graphical notation for quantum circuits equipped with a powerful set of rewrite rules that enable a graphical rewriting system for quantum computation. Given two quantum circuits U and U', they produce their corresponding representations as ZX-diagrams D and D'. These diagrams are then combined into $D^{\dagger}D'$ and simplified using the set of rewrite rules. If the result is in the form of the identity diagram, they can conclude their equivalence. Otherwise, nothing can be concluded because there are multiple forms for a ZX diagram in general. This approach is intuitive when we can see which rewrite rules are used and how ZX diagrams are changed accordingly. Our approach based on Dirac notation may be less intuitive, but it has never failed to conclude the equivalence of two quantum circuits, while their approach may fail to prove the equivalence of two equivalent quantum circuits.

8 Conclusion

We have presented an approach for checking the equivalence of quantum circuits based on Dirac notation. The equivalence checking process is simplified to comparing each column vector of two unitary matrices, representing two quantum circuits, modulo the same global phase. To eliminate the global phase during the comparison, we compare their corresponding density matrices instead of their column vectors. Moreover, our approach can take into account the constant inputs, making it more reliable in determining the equivalence problem compared to state-of-the-art tools in [3,17]. We have used Maude to develop the so-called $|QCEC\rangle$ tool for our approach based on the algorithm derived from our main theorem. Several case studies have been conducted with $|QCEC\rangle$, demonstrating the effectiveness of our approach and $|QCEC\rangle$. In addition to the future work discussed in the paper, we also aim to integrate our approach into the state-of-the-art tool presented in [3,24] to leverage its powerful simulation capabilities for quantum computation in checking the equivalence of quantum circuits.

References

1. Amy, M., Maslov, D., Mosca, M., Roetteler, M.: A meet-in-the-middle algorithm for fast synthesis of depth-optimal quantum circuits. Trans. Comp.-Aided Des. Integr. Circuits Syst. **32**(6), 818–830 (2013). https://doi.org/10.1109/TCAD.2013.2244643
2. Bennett, C.H., Wiesner, S.J.: Communication via one- and two-particle operators on Einstein-Podolsky-Rosen states. Phys. Rev. Lett. **69**, 2881–2884 (1992). https://doi.org/10.1103/PhysRevLett.69.2881
3. Burgholzer, L., Wille, R.: Advanced equivalence checking for quantum circuits. IEEE Trans. Comput. Aided Des. Integr. Circuits Syst. **40**(9), 1810–1824 (2021). https://doi.org/10.1109/TCAD.2020.3032630
4. Clavel, M., Durán, F., Eker, S., Lincoln, P., Martí-Oliet, N., Meseguer, J., Talcott, C.: All about Maude—a high-performance logical framework: how to specify, program and verify systems in rewriting logic. In: Lecture Notes in Computer Science, vol. 4350. Springer, Berlin (2007).https://doi.org/10.1007/978-3-540-71999-1
5. Cowtan, A., Dilkes, S., Duncan, R., Krajenbrink, A., Simmons, W., Sivarajah, S.: On the qubit routing problem. In: 14th Conference on the Theory of Quantum Computation, Communication and Cryptography, TQC 2019, 2019, University of Maryland, College Park, Maryland, USA. LIPIcs, vol. 135, pp. 5:1–5:32. Schloss Dagstuhl - Leibniz-Zentrum für Informatik (2019). https://doi.org/10.4230/LIPICS.TQC.2019.5
6. Dirac, P.A.M.: A new notation for quantum mechanics. Math. Proc. Cambridge Philos. Soc. **35**(3), 416–418 (1939). https://doi.org/10.1017/S0305004100021162
7. Do, C.M., Ogata, K.: Symbolic model checking quantum circuits in Maude. In: The 35th International Conference on Software Engineering and Knowledge Engineering, SEKE 2023, pp. 103–108 (2023). https://doi.org/10.18293/SEKE2023-014
8. Do, C.M., Ogata, K.: Theoretical foundation for equivalence checking of quantum circuits. In: The 2nd International Workshop on Formal Analysis and Verification of Post-Quantum Cryptographic Protocols, FAVPQC 2023 (2023). https://favpqc2023.gitlab.io/files/papers/6-Canh.pdf
9. Garcia-Escartin, J.C., Chamorro-Posada, P.: Equivalent Quantum Circuits (2011). https://doi.org/10.48550/arXiv.1110.2998
10. Giles, B., Selinger, P.: Exact synthesis of multiqubit Clifford+T circuits. Phys. Rev. A **87**, 032332 (2013). https://doi.org/10.1103/PhysRevA.87.032332
11. Maslov, D., Dueck, G.W., Miller, D.M., Negrevergne, C.: Quantum circuit simplification and level compaction. Trans. Comp.-Aided Des. Integr. Circuits Syst. **27**(3), 436–444 (008). https://doi.org/10.1109/TCAD.2007.911334
12. Maslov, D.: Advantages of using relative-phase Toffoli gates with an application to multiple control Toffoli optimization. Phys. Rev. A **93**(2) (2016). https://doi.org/10.1103/physreva.93.022311
13. Meseguer, J.: Twenty years of rewriting logic. J. Logic Algebraic Program. **81**(7), 721–781 (2012). https://doi.org/10.1016/j.jlap.2012.06.003, rewriting Logic and its Applications
14. Nam, Y., Ross, N.J., Su, Y., Childs, A.M., Maslov, D.: Automated optimization of large quantum circuits with continuous parameters. NPJ Quantum Inf. **4**(1), 23 (2018). https://doi.org/10.1038/s41534-018-0072-4
15. Nielsen, M.A., Chuang, I.L.: Quantum Computation and Quantum Information: 10th Anniversary Edition. Cambridge University Press (2010). https://doi.org/10.1017/CBO9780511976667

16. Paykin, J., Rand, R., Zdancewic, S.: Qwire: A core language for quantum circuits. SIGPLAN Not. **52**(1), 846–858 (2017). https://doi.org/10.1145/3093333.3009894
17. Peham, T., Burgholzer, L., Wille, R.: Equivalence checking of quantum circuits with the ZX-calculus. IEEE J. Emerging Sel. Top. Circuits Syst. **12**(3), 662–675 (2022).https://doi.org/10.1109/jetcas.2022.3202204
18. Sasanian, Z., Miller, D.M.: Reversible and quantum circuit optimization: A functional approach. In: Reversible Computation, pp. 112–124. Springer, Berlin (2013). https://doi.org/10.1007/978-3-642-36315-3_9
19. Shi, W., Cao, Q., Deng, Y., Jiang, H., Feng, Y.: Symbolic reasoning about quantum circuits in Coq. J. Comput. Sci. Technol. **36**(6), 1291–1306 (2021). https://doi.org/10.1007/s11390-021-1637-9
20. Shor, P.: Algorithms for quantum computation: discrete logarithms and factoring. In: Proceedings 35th Annual Symposium on Foundations of Computer Science, pp. 124–134 (1994). https://doi.org/10.1109/SFCS.1994.365700
21. Siraichi, M.Y., Santos, V.F.D., Collange, C., Pereira, F.M.Q.: Qubit allocation. In: Proceedings of the 2018 International Symposium on Code Generation and Optimization, pp. 113–125. CGO 2018, Association for Computing Machinery (2018). https://doi.org/10.1145/3168822
22. Takagi, T., Do, C.M., Ogata, K.: Automated quantum program verification in dynamic quantum logic. In: Dynamic Logic. New Trends and Applications, pp. 68–84. Springer Nature Switzerland (2024).https://doi.org/10.1007/978-3-031-51777-8_5
23. Wille, R., Burgholzer, L., Zulehner, A.: Mapping quantum circuits to IBM QX architectures using the minimal number of SWAP and H operations. In: Proceedings of the 56th Annual Design Automation Conference 2019, DAC 2019, p. 142. ACM, Las Vegas, NV, USA (2019). https://doi.org/10.1145/3316781.3317859
24. Wille, R., Hillmich, S., Burgholzer, L.: Tools for quantum computing based on decision diagrams. ACM Trans. Quantum Comput. **3**(3) (2022). https://doi.org/10.1145/3491246
25. Zulehner, A., Paler, A., Wille, R.: Efficient mapping of quantum circuits to the IBM QX architectures. In: 2018 Design, Automation & Test in Europe Conference and Exhibition (DATE), pp. 1135–1138 (2018). https://doi.org/10.23919/DATE.2018.8342181

Unified Opinion Dynamic Modeling as Concurrent Set Relations in Rewriting Logic

Carlos Olarte[1]($\boxtimes$), Carlos Ramírez[2], Camilo Rocha[2], and Frank Valencia[2,3]

[1] LIPN, CNRS UMR 7030, Université Sorbonne Paris Nord, Villetaneuse, France
olarte@lipn.univ-paris13.fr
[2] Department of Electronics and Computer Science, Pontificia Universidad Javeriana, Cali, Colombia
[3] CNRS-LIX, École Polytechnique de Paris, Palaiseau, France

Abstract. Social media platforms have played a key role in weaponizing the polarization of social, political, and democratic processes. This is, mainly, because they are a medium for opinion formation. Opinion dynamic models are a tool for understanding the role of specific social factors on the acceptance/rejection of opinions and they can be used to analyze certain assumptions on human behaviors. This work presents a framework that uses concurrent set relations as the formal basis to specify, simulate, and analyze social interaction systems with dynamic opinion models. Standard models for social learning are obtained as particular instances of the proposed framework. It has been implemented in the Maude system as a fully executable rewrite theory that can be used to better understand how opinions of a system of agents can be shaped. This paper also reports an initial exploration in Maude on the use of reachability analysis, probabilistic simulation, and statistical model checking of important properties related to opinion dynamic models.

Keywords: Concurrent set relations · Opinion dynamic models · Social interaction systems · Belief revision · Rewriting logic · Formal verification

1 Introduction

Social media platforms have played a key role in the polarization of social, political, and democratic processes. Social uprisings in the Middle East, Asia, and Central and South America have led to sudden changes in the structure and nature of society during this past decade [6,12,15,19,20,30]. Polarization across the globe has paved the way to the divergence of political attitudes away from

---

This work has been partially supported by the SGR project PROMUEVA (BPIN 2021000100160) under the supervision of Minciencias (Ministerio de Ciencia Tecnología e Innovación, Colombia).

the center, towards ideological extremes, sometimes resulting in fractured institutions, erratic policy making, incipient political dialog, and the resurgence of old discredited regimes [11,13,16,18,20,22]. Democracy, viewed as a system of power controlled by the people, has been made vulnerable by severe polarization as opposing sides are seen as adversaries that compete against an enemy needing to be vanquished. As a result, popular election campaigns—including presidential ones—have compromised the basic principles of democratic election in some countries [3,4,28,29]. All these scenarios have a common factor: social media interaction as a medium for *opinion formation* fueling polarization.

Social learning and opinion dynamic models have been developed to understand the role of specific social factors on the acceptance/rejection of opinions, such as the ones communicated via social media (see, e.g., [2,8,14,17]). They are often used to validate how certain assumptions on human behaviors can explain alternative scenarios, such as opinion consensus, polarization, and fragmentation. In their micro-level approach, the one followed in the present work, users are considered as agents that can share opinions on a given topic. They update their opinion by interacting with a selected group of users that have some influence on them (e.g., influencers, their family and friends). These dynamics take place at discrete time steps at which (some) agents update their opinion. For instance, an opinion model can deterministically update the opinion of all agents in such a time-step, while another one can non-deterministically update the opinion of a single agent. Depending on the model of choice, which usually defines its own update function for the individual agents, phenomena under different assumptions can be observed. The ultimate goal is to understand how the opinions of the agents, as a social system, are shaped after a certain number of steps.

This work proposes a framework that uses concurrent set relations as the formal basis to specify, simulate, and analyze social interaction systems with dynamic opinion models. The framework uses *influence graphs* to specify the structure of agent interactions in the social system under study: vertices represent agents and a directed weighted edge from a to b represents the weighted influence of agent's a opinion over the opinion of agent b. In the sense of set relations in [25], the framework comprises two main mechanisms that are combined via closures for specifying opinion dynamics over the graphs: namely, an atomic set relation and a strategy. The *atomic set relation* updates the opinion of a single vertex with respect to a set of edges (and the corresponding vertices) incident to it. The *strategy* selects the edges that will be used to update in parallel (i.e., synchronously) the opinion associated to the vertices with edges incident to it in the given set. As a consequence, dynamic opinion models can be formalized as a concurrent set relation system, with parametric update function, using the composition of an atomic relation and a strategy via closures. An important observation is that the determinism or non-determinism inherent to a given opinion dynamic model is exactly captured by the deterministic or non-deterministic nature of the corresponding concurrent set relation.

Standard models for social learning are obtained as particular instances of the proposed framework. The classical DeGroot opinion model [9] is obtained as the synchronous closure under the maximal redices strategy of a given atomic set relation. In a similar fashion, gossip-based models that use pairwise interactions to represent the opinion formation process (see, e.g., [10]) are obtained via the asynchronous closure where the strategy selects single edges for the given atomic set relation. Other opinion models can be obtained via the synchronous closure of an atomic set relation, as midpoints between De Groot and gossip-based models.

The proposed framework is implemented in the Maude system [7]. It is a rewrite theory that exploits the reflective capabilities of rewriting logic and that can be instantiated to the opinion model of interest. A state is an object-like configuration representing the structure of the system and its opinion values. An object is either an agent u with its opinion o_u, specified as $\langle u : o_u \rangle$, or the influence of agent u over agent v with weight i_{uv}, specified as $\langle (u,v) : i_{uv} \rangle$. The update function μ of each specific model is to be defined equationally. The implementation of both the atomic set relation and the strategy is inspired by the ideas in [24]. The atomic set relation is axiomatized as a (non-executable) rewrite rule that takes as input an agent $\langle u : o_u \rangle$ and a set of edges $A \subseteq E$ in the current state. For a given state, it updates the opinion o_u to a new opinion o'_u using μ, and the opinion and influence of agents adjacent to it w.r.t. A. As a result, each atomic step rewrites a single object $\langle u : o_u \rangle$ to its updated version $\langle u : o'_u \rangle$. The metalevel is used to apply the atomic rewrite rule over the agents in a state according to the edges selected by the given strategy: only agents appearing as targets of the directed edges can have their opinion updated. This strategy is defined equationally by the user and computes a collection of subsets of E: a parallel rewrite step under the maximal redices strategy is performed for each subset A of edges. Since the atomic rewrite relation is deterministic, the strategy is the only source of non-determinism in the system and a concurrent step is made for each identified subset A. The implementation of the proposed framework results in a fully executable object-like rewrite theory in Maude. This tool can be used to better understand how opinions of a system of agents are shaped—and to ultimately understand polarization—using formal methods techniques, such as reachability analysis and temporal model checking.

This work is part of a broader effort to make available computational ideas and approaches for analyzing phenomena in social networks, such as polarization, consensus, and fragmentation. They include concurrency models, modal and probabilistic logics, and formal methods frameworks, techniques, and tools. In this context, the work presented here is a first step towards the use of rewriting logic for such purposes. As explained in the sections that follow, one major problem a opinion dynamic model may face is that of state space explosion. An initial exploration on the use of probabilistic simulation and statistical analysis to deal with this problem is reported in this work. However, the extension of the proposed framework to a fully probabilistic setting, in which—e.g.—the strategy selects the set of edges according to a probability distribution function, falls outside the scope of this paper. It needs to be further explored as future work as it may open the door to statistical model checking of novel properties using a new breed of measures, thus paving the way to the analysis of quantitative properties beyond the reach of techniques currently available for opinion dynamic models.

Organization. After recalling the notion of set relations in Sects. 2 and 3 shows how different models for social learning can be seen as particular instances (atomic set relation and strategy) of this framework. The implementation in Maude is described in Sect. 4, while different analyses performed on the proposed rewrite theory are introduced in Sect. 5. Section 6 concludes the paper. The full Maude specification supporting the set relations framework is available at [23], as companion tool to the paper.

2 Set Relations

This section introduces set relations and their notation, as used in this paper. It defines the asynchronous, parallel, and synchronous set relations as closures of an atomic set relation. This section is based mainly on [25].

Let $\mathcal{U}$ be a set whose elements are denoted $A, B, \ldots$ and let $\rightarrow$ be a binary relation on $\mathcal{U}$. An element A of $\mathcal{U}$ is called a $\rightarrow$-*redex* iff there exists $B \in \mathcal{U}$ such that the *pair* $\langle A; B \rangle \in \rightarrow$. The expressions $A \rightarrow B$ and $A \not\rightarrow B$ denote $\langle A; B \rangle \in \rightarrow$ and $\langle A; B \rangle \notin \rightarrow$, respectively. The *identity* and *reflexive-transitive* closures of $\rightarrow$ are defined as usual and denoted $\xrightarrow{0}$ and $\xrightarrow{*}$, respectively.

It is assumed that $\mathcal{U}$ is the family of all *nonempty* finite subsets of an abstract and possibly infinite set T whose members are called *elements* (i.e., $\mathcal{U} \subseteq \mathcal{P}(T)$, $\emptyset \notin \mathcal{U}$, and if $A \in \mathcal{U}$, then $\text{card}(A) \in \mathbb{N}$). Therefore, $\rightarrow$ is a binary relation on finite subsets of elements in T. When it is clear from the context, curly brackets are omitted from set notation; e.g., $a, b \rightarrow b$ denotes $\{a, b\} \rightarrow \{b\}$. Because this convention, the symbol ',' is overloaded to denote set union. For example, if A denotes the set $\{a, b\}$, B the set $\{c, d\}$, and D the set $\{d, e\}$, the expression $A, B \rightarrow B, D$ denotes $a, b, c, d \rightarrow c, d, e$.

Given a set of elements, in the asynchronous set relation *exactly one* redex is selected to be updated.

Definition 1. (*Asynchronous Set Relation*) The *asynchronous relation* $\xrightarrow{\square}$ is defined as the asynchronous closure of $\rightarrow$, i.e., the set of pairs $\langle A; B \rangle \in \mathcal{U} \times \mathcal{U}$ such that $A \xrightarrow{\square} B$ iff there exists a $\rightarrow$-redex $A' \subseteq A$ and an element $B' \in \mathcal{U}$ such that $A' \rightarrow B'$ and $B = (A \setminus A') \cup B'$.

In the parallel set relation, a nonempty collection of redices is identified to be updated in parallel (i.e., without interleaving).

Definition 2. (*Parallel Set Relation*) The *parallel relation* $\xrightarrow{\parallel}$ is defined as the parallel closure of $\rightarrow$, i.e., the set of pairs $\langle A; B \rangle \in \mathcal{U} \times \mathcal{U}$ such that $A \xrightarrow{\parallel} B$ iff there exist (nonempty) pairwise disjoint $\rightarrow$-redices $A_1, \ldots, A_n \subseteq A$, and elements $B_1, \ldots, B_n$ in $\mathcal{U}$ such that $A_i \rightarrow B_i$, for $1 \leq i \leq n$, and $B = \left(A \setminus \bigcup_{1 \leq i \leq n} A_i\right) \cup \left(\bigcup_{1 \leq i \leq n} B_i\right)$.

The synchronous set relation $\xrightarrow{s}$ applies as many atomic reductions as possible, in parallel. However, in contrast to the previous two closures, the redices are selected with the help of a *strategy* s, namely, a function that identifies a nonempty subset of redices. As a consequence, the synchronous set relation is a subset of the parallel set relation. It is important to note that the notion of strategy used for defining the synchronous closure of the atomic set relation is different to the one introduced in Sect. 1 for the framework; the name used in this section is kept from [25].

Definition 3. ($\rightarrow$-*strategy*) A $\rightarrow$-*strategy* is a function s that maps any element $A \in \mathcal{U}$ into a set $s(A) \subseteq \mathcal{P}(\rightarrow)$ such that if $s(A) = \{\langle A_1; B_1 \rangle, \ldots \langle A_n; B_n \rangle\}$, then $A_i \subseteq A$ and $A_i \rightarrow B_i$, for $1 \leq i \leq n$, and $A_1, \ldots, A_n$ are pairwise disjoint.

Definition 4. (*Synchronous Relation*) Let s be a $\rightarrow$-strategy. The *synchronous relation* $\xrightarrow{s}$ is defined as the synchronous closure of $\rightarrow$ w.r.t. s, i.e., the set of pairs $\langle A; B \rangle \in \mathcal{U} \times \mathcal{U}$ such that $A \xrightarrow{s} B$ iff $B = \left(A \setminus \bigcup_{1 \leq i \leq n} A_i \right) \cup \left(\bigcup_{1 \leq i \leq n} B_i \right)$ where $s(A) = \{\langle A_1; B_1 \rangle, \ldots \langle A_n; B_n \rangle\}$.

This section is concluded with an example that illustrates the notions introduced so far.

Vaccine Example. Consider the directed weighted graph $G = (V, E, i)$ in Fig. 1. It represents a social system with six agents $V = \{a, b, c, d, e, f\}$ and twelve opinion influences. The label $i(u, v)$ associated to each edge (u, v) from agent u to agent v denotes the opinion influence $i_{uv} = i(u, v)$ of agent u over the opinion of agent v (about a given topic): these values are in the real interval $[0, 1]$ (i.e., $i : E \rightarrow [0, 1]$); the higher the value, the stronger the influence. In this example, the influence of f over a is the strongest possible. Notice that agents may also have *self-influence*, representing agents whose opinion need not be completely influenced by the opinion of the others.

The initial opinions (or beliefs) of the agents are depicted within the box below each node. They are specified by a function $o : V \rightarrow [0, 1]$, which is assumed to represent the opinion value $o_u = o(u)$ of each agent u on the given topic. The greater the value, the stronger (weaker) the agreement (disagreement) with the proposition, and 0 represents total disagreement. In this example such a proposition is *vaccines are safe*. Intuitively, the agents a, b, and c are in strong disagreement with vaccines being safe (the anti-vaxxers) and the rest are in strong agreement (the pro-vaxxers).

Notice that although a is the most extreme anti-vaxxer, the most extreme pro-vaxxer f has a strong influence over a. Hence, it is expected that the evolution of a's opinion will be highly influenced by the opinion of f. In general, an agent's opinion evolution takes into account a subset of its influences, as will be explained shortly.

Recall the object-like notation in Sect. 1. The set of elements T is made of pairs of the form $\langle u : r \rangle$ or $\langle (u, v) : r \rangle$, with $u, v \in V$, $(u, v) \in E$, and $r \in [0, 1]$.

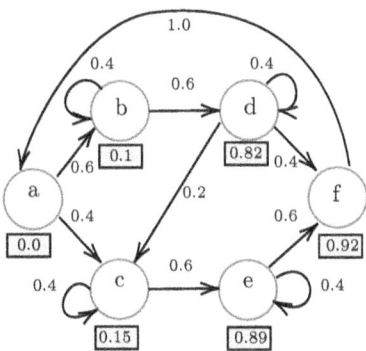

Fig. 1. Graph representing opinion and influence interaction in a social system. Initial opinions are given within the box below each node. The labels on each edge (u, v) represent the influence value of agent u over agent v

The graph in Fig. 1 can be specified as the set of elements Γ:

$\Gamma = \{\langle a : 0.0\rangle, \langle b : 0.1\rangle, \langle c : 0.15\rangle, \langle d : 0.82\rangle, \langle e : 0.89\rangle, \langle f : 0.92\rangle,$
$\langle (a,b) : 0.6\rangle, \langle (a,c) : 0.4\rangle, \langle (b,d) : 0.6\rangle, \langle (c,e) : 0.6\rangle, \langle (d,c) : 0.2\rangle, \langle (d,f) : 0.4\rangle,$
$\langle (e,f) : 0.6\rangle, \langle (f,a) : 1.0\rangle, \langle (b,b) : 0.4\rangle, \langle (c,c) : 0.4\rangle, \langle (d,d) : 0.4\rangle, \langle (e,e) : 0.4\rangle\}.$

The atomic relation $\to_A$ is defined over elements representing agents and is parametric on a set A of elements representing edges in Γ. In this example, it follows the pattern

$$\langle u : o_u\rangle \to_A \langle u : \sum_{\langle (x,u):i_{xu}\rangle \in A} o_x \cdot \frac{i_{xu}}{\sum_{\langle (y,u):i_{yu}\rangle \in A} i_{yu}}\rangle, \quad (1)$$

where the summation in the denominator is assumed to be non-zero. The opinion o_u of an agent u w.r.t. to A is updated to be the weighted average of the opinion values of those agents adjacent to u and whose influence is present in A. For instance, let $A = \{\langle (a,b) : 0.6\rangle, \langle (b,b) : 0.4\rangle, \langle (c,e) : 0.6\rangle\}$. Then, the atomic set relation $\to_A$ has the following two pairs:

$$\langle b : 0.1\rangle \to_A \langle b : 0.04\rangle \qquad \langle e : 0.89\rangle \to_A \langle e : 0.15\rangle.$$

In the case of agent b, its opinion is updated to $0.04 = 0.0 \cdot \frac{0.6}{1.0} + 0.1 \cdot \frac{0.4}{1.0}$ because, w.r.t. A, it is influenced both by itself and by agent a, whose opinion value is 0.0 and influence over b is 0.6. In the case of agent e, its opinion is influenced only by agent c. In this case, the value is updated to $0.15 = 0.15 \cdot \frac{0.6}{0.6}$. It can be said that, w.r.t. A, agent e acts like a *puppet* whose own opinion is not taken into account when it is updated.

The asynchronous closure of $\to_A$ has exactly two pairs, one for each redex determined by $\to_A$ (i.e., one for agent b and another for agent e):

$$\Gamma \xrightarrow{\square}_A (\Gamma \setminus \{\langle b : 0.1\rangle\}) \cup \{\langle b : 0.04\rangle\} \qquad \Gamma \xrightarrow{\square}_A (\Gamma \setminus \{\langle e : 0.89\rangle\}) \cup \{\langle e : 0.15\rangle\}.$$

The parallel closure $\xrightarrow{\parallel}$ has three pairs: one in which the opinions of both b and e are updated, in addition to the same two pairs present in the asynchronous closure:

$$\Gamma \xrightarrow{\parallel}_A (\Gamma \setminus \{\langle b : 0.1\rangle\}) \cup \{\langle b : 0.04\rangle\}$$
$$\Gamma \xrightarrow{\parallel}_A (\Gamma \setminus \{\langle e : 0.89\rangle\}) \cup \{\langle e : 0.15\rangle\}$$
$$\Gamma \xrightarrow{\parallel}_A (\Gamma \setminus \{\langle b : 0.1\rangle, \langle e : 0.89\rangle\}) \cup \{\langle b : 0.04\rangle, \langle e : 0.15\rangle\}.$$

Finally, to illustrate the synchronous closure of $\to$, let $s = A$ be the strategy. That is, all redices in $\to_A$ are identified to be reduced. Therefore, this relation has the only pair in which the opinions of both b and e are updated in parallel:

$$\Gamma \xrightarrow{s}_A (\Gamma \setminus \{\langle b : 0.1\rangle, \langle e : 0.89\rangle\}) \cup \{\langle b : 0.04\rangle, \langle e : 0.15\rangle\}.$$

3 Opinion Dynamic Models

This section shows how opinion dynamic models can be specified as set relations (see Sect. 2). In particular, a gossip-based and the classical De Groot opinion models are introduced, as well as a generalization of De Groot and gossip (under some conditions), here called the *hybrid opinion model*.

The three above-mentioned models are defined, as stated in Sect. 2, over a directed weighted graph $G = (V, E, i)$ representing a social system, with agents V, directed opinion influences $E \subseteq V \times V$, and influence values $i : E \to [0, 1]$. A given topic (i.e., proposition) is fixed. The weight $i_{uv} = i(u, v)$ associated to each edge $(u, v) \in E$ from agent u to agent v denotes the opinion influence value of agent u over the opinion value of agent v on the given topic. The opinion value $o_u = o(u) \in [0, 1]$ associated to each agent $u \in V$ in the given topic is assumed to be known by all agents in the system. As in Sect. 2, the higher the value of a opinion (resp., influence), the stronger the agreement (resp., influence).

The set of elements T in the set relations framework is made of pairs of the form $\langle u : r \rangle$ or $\langle (u, v) : r \rangle$, with $u, v \in V$, $(u, v) \in E$, and $r \in [0, 1]$. A G-*configuration* (or *configuration*) is the set of elements in T that exactly represent the structure of G, and the values of opinions and influences. Therefore, in the rest of this section, it is assumed that any configuration Γ can be partitioned in two sets Γ_o and Γ_i, respectively containing elements of the form $\langle u : o_u \rangle$ specifying opinions and $\langle (u, v) : i_{uv} \rangle$ specifying influences.

A model specifies how opinions (associated to agents) can be updated. Each model definition comprises three pieces; namely, an atomic relation, a strategy, and an update function for opinions. Therefore, a model specifies how a G-configuration $\Gamma = \Gamma_o \cup \Gamma_i$ can change to another G-configuration $\Gamma' = \Gamma_{o'} \cup \Gamma_i$, where only opinions are updated. It is important to note that the notion of strategy introduced in this section generalizes the notion of strategy introduced in Sect. 2, as will be explained later.

The atomic relation is defined in Sect. 3.1 for the three models. Each model is introduced by identifying a specific strategy and a specific update function in subsequent sections.

3.1 The Atomic Relation

The atomic relation $\to_A$ is parametric on a subset $A \subseteq \Gamma_i$ and defines how the opinion of a single agent may evolve. The set of influences A directly identifies the influences (and indirectly the opinions) to update the opinion of each agent in the configuration Γ (i.e., in Γ_o). For each one of the three models, the atomic relation $\to_A$ follows the pattern:

$$\langle u : o_u \rangle \to_A \langle u : \mu(\Gamma, A, u) \rangle, \qquad (A-Rel)$$

where μ is the update function specific to each model. This function takes as input a G-configuration (e.g., Γ), a subset of its influences (e.g., A), and the agent whose opinion is to be updated (e.g., u), and outputs the new opinion for agent u w.r.t. Γ and A in the corresponding model.

3.2 Gossip-Based Models

In a gossip-based model, single peer-to-peer interactions are used to update the opinion of a single user at each time-step. In general, a strategy in the proposed framework identifies a collection of subsets of interactions in Γ_i. In particular, the strategy ρ_{gossip} maps a G-configuration to the collection of singletons made from the influences in Γ_i:

$$\rho_{\text{gossip}}(\Gamma) = \{\{x\} \mid x \in \Gamma_i\}.$$

This means that, at each time-step, the opinion value of agent v can be updated w.r.t. the opinion value of agent u for each singleton $\{\langle (u,v) : i_{uv} \rangle\}$ computed by the strategy $\rho_{\text{gossip}}(\Gamma)$.

The update function μ_{gossip} is defined for any $u \in V$ and $A = \{\langle (v,u) : i_{vu}\rangle\} \in \rho_{\text{gossip}}(\Gamma)$ as:

$$\mu_{\text{gossip}}(\Gamma, A, u) = o_u + (o_v - o_u) \cdot i_{vu}.$$

Each singleton $A \in \rho_{\text{gossip}}(\Gamma)$ determines an atomic relation that updates exactly one agent's opinion in the given configuration. Recall, from Sect. 3.1, that each pair in the atomic set relation $\to_A$ has the form:

$$\langle u : o_u \rangle \to_A \langle u : \mu_{\text{gossip}}(\Gamma, A, u) \rangle.$$

Hence, in this model, the opinion of an agent u is updated by identifying an edge from an agent v (it may be u itself if it has a self-loop) with influence i_{vu} over u and by adding to its current opinion o_u the weighted difference of opinion $(o_v - o_u) \cdot i_{vu}$ of v over u.

A gossip-based model is identified as a binary set relation on G-configurations in terms of the asynchronous closure of $\to_A$, for each singleton $A \in \rho_{\text{gossip}}(\Gamma)$.

Definition 5. The $\to_{\text{gossip}}$ set relation is defined as the set of pairs $\langle \Gamma; \Gamma' \rangle$ of G-configurations such that:

$$\Gamma \to_{\text{gossip}} \Gamma' \quad \text{iff} \quad (\exists A \in \rho_{\text{gossip}}(\Gamma)) \; \Gamma \xrightarrow{\square}_A \Gamma'.$$

From the viewpoint of concurrency, the gossip-based opinion dynamic model captured by $\to_{\text{gossip}}$ is non-deterministic in the sense that at each state (i.e., G-configuration) exactly $|\Gamma_i|$ transitions are possible, one per edge in E.

3.3 De Groot Model

In the De Groot model, the opinion value of every agent in the network is updated at each time-step. All influences are considered at the same time.

The strategy for De Groot in the proposed framework identifies the whole set of interactions in the network, i.e., Γ_i. In particular, the strategy ρ_{DeGroot} maps a G-configuration to the singleton whose only element is Γ_i:

$$\rho_{\text{DeGroot}}(\Gamma) = \{\Gamma_i\}.$$

The update function μ_{DeGroot} is defined for any $u \in V$ and $A \in \rho_{\text{DeGroot}}(\Gamma)$ (i.e., $A = \Gamma_i$) as:

$$\mu_{\text{DeGroot}}(\Gamma, A, u) = o_u + \sum_{\langle (v,u):i_{vu}\rangle \in A} (o_v - o_u) \cdot \frac{i_{vu}}{\sum_{\langle (x,u):i_{xu}\rangle \in A} i_{xu}},$$

where the summation in the denominator is assumed to be non-zero. Otherwise, the value of this function is assumed to be o_u (i.e., the opinion of agent u does not change).

The De Groot model is identified as a binary set relation on G-configurations in terms of the synchronous closure of $\to_{\Gamma_i}$ under the maximal redices strategy for $s = \Gamma_i$.

Definition 6. The $\to_{\text{DeGroot}}$ set relation is defined as the set of pairs $\langle \Gamma; \Gamma' \rangle$ of G-configurations such that:

$$\Gamma \to_{\text{DeGroot}} \Gamma' \quad \text{iff} \quad \Gamma \xrightarrow{\Gamma_i}_{\Gamma_i} \Gamma'.$$

From the viewpoint of concurrency, the De Groot opinion dynamic model captured by $\to_{\text{DeGroot}}$ is deterministic in the sense that, at each state, there is exactly only one possible transition where all influences are taken into account to update each agent's opinion without interleaving.

3.4 The Hybrid Model

The hybrid model considers every possible influence scenario in the network, i.e., any possible combination of influences are used to update the opinion of agents that may be affected by them at each time-step. Therefore, the strategy in the proposed framework identifies all nonempty subsets of interactions in Γ_i. In particular, the strategy ρ_{hybrid} maps a G-configuration to the collection of nonempty subsets made from the influences in Γ_i:

$$\rho_{\text{hybrid}}(\Gamma) = \{A \mid A \subseteq \Gamma_i \text{ and } A \neq \emptyset\}.$$

This means that, at each time-step, the opinion value of an agent v can be updated with a subset of its influencers.

The update function μ_{hybrid} is the same as function ρ_{DeGroot}. That is, it is defined for any $u \in V$ and $A \in \rho_{\text{hybrid}}(\Gamma)$ as:

$$\mu_{\text{hybrid}}(\Gamma, A, u) = o_u + \sum_{\langle(v,u):i_{vu}\rangle \in A} (o_v - o_u) \cdot \frac{i_{vu}}{\sum_{\langle(x,u):i_{xu}\rangle \in A} i_{xu}},$$

where the summation in the denominator is assumed to be non-zero. Otherwise, the value of this function is assumed to be o_u (i.e., the opinion of agent u does not change). Each subset $A \in \rho_{\text{hybrid}}(\Gamma)$ determines an atomic relation that may update more that one agent's opinion. Hence, in this model, the opinion of an agent is updated by identifying some edges that may have influence over it.

The hybrid model is identified as a binary set relation on G-configurations in terms of the synchronous closure of $\to_A$, for each subset $A \in \rho_{\text{hybrid}}(\Gamma)$.

Definition 7. The $\to_{\text{hybrid}}$ set relation is defined as the set of pairs $\langle \Gamma; \Gamma' \rangle$ of G-configurations such that:

$$\Gamma \to_{\text{hybrid}} \Gamma' \quad \text{iff} \quad (\exists A \in \rho_{\text{hybrid}}(\Gamma))\ \Gamma \xrightarrow{A}_A \Gamma'.$$

From the viewpoint of concurrency, the hybrid opinion dynamic model has the maximum degree of non-determinism possible. Moreover, this model is more general than the De Groot model.

Theorem 1. $\to_{DeGroot} \subseteq \to_{hybrid}$.

Proof. It follows by noting that $\Gamma_i \in \rho_{\text{hybrid}}(\Gamma)$ and, for each vertex $u \in V$, the equality $\mu_{\text{DeGroot}}(\Gamma, \Gamma_i, u) = \mu_{\text{hybrid}}(\Gamma, \Gamma_i, u)$ holds.

It is not necessarily the case that $\to_{\text{gossip}} \subseteq \to_{\text{hybrid}}$. This is because the update functions do not always agree when the collection of selected influences A is a singleton. In particular, for each singleton $A = \{\langle(v,u) : i_{vu}\rangle\}$, $\mu_{\text{hybrid}}(\Gamma, A, u) = o_v$, meaning that agent u in the hybrid model behaves always like a *puppet* when $u \neq v$. Note that this is not (necessarily) the case in $\to_{\text{gossip}}$. Nevertheless, there is a class of graphs for which this inclusion holds.

Theorem 2. *If G is such that each vertex has a self-loop and is influenced at most by another vertex, and the summation of its incoming influences is 1, then* $\to_{gossip} \subseteq \to_{hybrid}$.

Proof. If $\Gamma \to_{\text{gossip}} \Gamma'$, there is a singleton $A \in \rho_{\text{gossip}}(\Gamma)$ such that $\Gamma \xrightarrow{\square}_A \Gamma'$. Let $A = \{\langle(v,u) : i_{vu}\rangle\}$. If u has exactly one incoming edge, then $v = u$ (by the initial assumption) and $\rho_{\text{gossip}}(\Gamma, A, u) = o_u = \rho_{\text{hybrid}}(\Gamma, A, u)$. Since $A \in \rho_{\text{hybrid}}(\Gamma)$, it follows that $\Gamma \to_{\text{hybrid}} \Gamma'$. If u has two edges, and the self-loop is taken, the case $v = u$ is as above. Otherwise, if $u \neq v$, the same transition is obtained in the hybrid model by taking $A' \in \rho_{\text{hybrid}}(\Gamma)$ where $A' = A \cup \{\langle(u,u) : 1 - i_{vu}\rangle\}$ (an noticing that the denominator in μ_{hybrid} becomes 1).

4 The Framework in Rewriting Logic

This section presents a rewrite theory that implements the set relations framework in Sect. 2. Off-the-shelf definitions are provided to instantiate the framework with opinion dynamic models, such as the ones introduced in Sect. 3. This section assumes familiarity with rewriting logic [21] and Maude [7]; Sect. 4.1 presents some preliminaries on these two subjects. The full Maude specification supporting the set relations framework is available at [23].

4.1 Overview of Rewriting Logic and Maude

A *rewrite theory* [21] is a tuple $\mathcal{R} = (\Sigma, E, L, R)$ such that: (Σ, E) is an equational theory where Σ is a signature that declares sorts, subsorts, and function symbols; E is a set of (conditional) equations of the form $t = t'\,\mathbf{if}\,\psi$, where t and t' are terms of the same sort, and ψ is a conjunction of equations; L is a set of *labels*; and R is a set of labeled (conditional) rewrite rules of the form $l : q \longrightarrow r\,\mathbf{if}\,\psi$, where $l \in L$ is a label, q and r are terms of the same sort, and ψ is a conjunction of equations. Condition ψ in equations and rewrite rules can be more general than conjunction of equations, but this extra expressiveness is not needed in this paper.

The expression $T_{\Sigma,s}$ denotes the set of ground terms of sort s and $T_\Sigma(X)_s$ denotes the set of terms of sort s over a set of sorted variables X. The expressions $T_\Sigma(X)$ and T_Σ denote all terms and ground terms, respectively. A substitution $\sigma : X \to T_\Sigma(X)$ maps each variable to a term of the same sort and $t\sigma$ denotes the term obtained by simultaneously replacing each variable x in a term t with $\sigma(x)$.

A *one-step rewrite* $t \longrightarrow_\mathcal{R} t'$ holds if there is a rule $l : q \longrightarrow r\,\mathbf{if}\,\psi$, a subterm u of t, and a substitution σ such that $u = q\sigma$ (modulo equations), t' is the term obtained from t by replacing u with $r\sigma$, and $v\sigma = v'\sigma$ holds in (Σ, E) for each $v = v'$ in ψ. The reflexive-transitive closure of $\longrightarrow_\mathcal{R}$ is denoted as $\longrightarrow^*_\mathcal{R}$.

Maude [7] is a language and tool supporting the specification and analysis of rewrite theories. A Maude module (mod M is ... endm) specifies a rewrite theory $\mathcal{R}$. Sorts and subsort relations are declared by the keywords sort and subsort; function symbols, or *operators*, are introduced with the op keyword: op $f : s_1 \ldots s_n$ -> s, where $s_1, \ldots, s_n$ are the sorts of its arguments, and s is its (value) sort. Operators can have user-definable syntax, with underbars '_' marking each of the argument positions (e.g., _+_). Some operators can have equational attributes, such as assoc, comm, and id: t, stating that the operator is, respectively, associative, commutative, and/or has identity element t. Equations are specified with the syntax eq $t = t'$ or ceq $t = t'$ if ψ; and rewrite rules as rl [l] : u => v or crl [l] : u => t' if ψ. The mathematical variables in such statements are declared with the keywords var and vars.

Maude provides a large set of analysis methods, including computing the canonical form of a term t (command red t), simulation by rewriting (rew t), reachability analysis (search t =>* t' such that ψ), and rewriting according to a given rewrite strategy (srew t using str). Basic rewrite strategies include $r[\sigma]$ (apply rule with label r once with the optional ground substitution σ), idle (identity), fail (empty set), and match P s.t. C, which checks whether the current term matches the pattern P subject to the constraint C. Compound strategies can be defined using concatenation ($\alpha\,;\,\beta$), disjunction ($\alpha\,|\,\beta$), iteration ($\alpha*$), α or-else β (execute β if α fails), among other options.

The Unified Maude model-checking tool [26] (umaudemc) enables the use of different model checkers to analyze Maude specifications. Besides being an interface for the standard LTL model checker of Maude, it also offers the possibility of interfacing external CTL and probabilistic model checkers. For the purpose of this paper, the command scheck [27] is used to assign probabilities to the transition system generated by an initial term t and to perform statistical model checking to estimate quantitive expressions written in the Quantitative Temporal Expressions (QuaTEx) language [1]. QuaTEx supports parameterized recursive temporal operator definitions using primitive non-temporal operators (e.g., conditional statements, values from the current state of the system, etc.) and the *next* temporal operator (notation #). The QuaTEx query eval E[*expr*] returns the expected value of the expression *expr* using the Monte Carlo method.

Meta-programming. Maude supports *meta-programming*, where a Maude module M (resp., a term t) can be (meta-)represented as a Maude *term* $\overline{M}$ of sort Module (resp., as a Maude term $\overline{t}$ of sort Term) in Maude's META-LEVEL module. Maude provides built-in functions such as metaRewrite and metaSearch, which are the "meta-level" functions corresponding to "user-level" commands to perform rewriting and search, respectively.

4.2 Influences, Opinions, and States

An agent a and its opinion o_a, and the influence of agent a over agent b with weight i_{ab}, are specified in $\mathcal{R}$ with the help of the following sorts and function symbols:

```
sorts Agent Opinion Edge .
op <_:_>       :   Agent       Float -> Opinion [ctor] .
op <'(_,_'):_> : Agent Agent Float -> Edge    [ctor] .
```

The user is expected to provide appropriate constructors for the sort Agent, e.g., by extending $\mathcal{R}$ with the subsort relation subsort Nat < Agent to use natural numbers as identifiers for agents.

Sets of agents, opinions, and edges (sorts SetAgent, SetOpinion, and SetEdge respectively) are defined as ","-separated sets of elements in the usual way. A G-configuration $\varGamma = \varGamma_o \cup \varGamma_i$ is represented by a term of sort Network, defining the set of agents' opinions ($\varGamma_o$) and influences ($\varGamma_i$) with the following sort and function symbol:

```
sort Network .
op < nodes:_ ; edges:_ > : SetOpinion SetEdge -> Network [ctor] .
```

Analyzing opinion dynamics usually requires determining the number of interactions between agents and the time needed to reach a given state. A term of the form "N in step: t comm: nc" of sort State represents the state of a network N at time instant t, where a number of interactions/communications nc have taken place:

```
sort State .
op _in step:_ comm:_ : Network Nat Nat -> State [ctor] .
```

4.3 Strategies and the Atomic Relation

The framework is parametric on a strategy ρ and an update function μ, as explained in Sect. 3. The atomic relation $\rightarrow_A$ is parametric on a nonempty subset $A \subseteq \Gamma_i$. A strategy identifies each one of such subsets at each time-step. A SetSetEdge is a ";"-separated set of sets of edges.

```
sort SetSetEdge . subsort NeSetEdge < SetSetEdge .
op mt :  -> SetSetEdge [ctor] .
op _;_ : SetSetEdge SetSetEdge -> SetSetEdge [ctor assoc comm id: mt] .
```

Some distinguished SetSetEdges include the singleton with all the edges in the network (De Groot model), the set containing only singletons (Gossip model) and the set of nonempty subsets of edges (Hybrid model).

```
var SE : SetEdge .   var E : Edge .
op deGroot  : SetEdge     -> SetSetEdge .
eq deGroot(SE) = SE .

op gossip : SetEdge       -> SetSetEdge .
eq gossip(empty) = mt .
eq gossip((E, SE)) = E ; gossip(SE) .

op hybrid : SetEdge -> SetSetEdge .
eq hybrid(SE) = power-set(SE) \ empty .

op strategy :  -> SetSetEdge . --- user defined strategy
```

The operator strategy must be defined by the user to identify the subsets $A \subseteq \Gamma_i$ available in each transition. This can be done, e.g., by adding the equation

```
eq strategy =  gossip(edges) .
```

where edges is the set of edges in the network currently being modeled.

The atomic relation (pattern (A-Rel)) is defined as a non-executable rewrite rule and the set relation framework is implemented using the meta-programming facilities in Maude. In particular, the atomic rewrite relation updates the BELIEF of a given AGENT (u in pattern (A-Rel)) to a new BELIEF' when a set of EDGES (A) is selected and the current state of the system is STATE (Γ):

```
var AGENT : Agent . vars BELIEF BELIEF' : Float . var STATE : State .
vars SETEDGE EDGES : SetEdge .

op update : State SetEdge Agent -> Float .   --- user defined μ

crl [atomic] : < AGENT : BELIEF >  =>  < AGENT :   BELIEF' >
    if BELIEF' := update(STATE, SETEDGE, AGENT) [nonexec] .
```

The function `update` (μ in pattern (A-Rel)) must be specified by the user. The framework provides instances of this function for the models presented in Sect. 3.

An asynchronous, parallel, or synchronous rewrite step, depending on the underlying strategy, is captured by the rewrite rule `step` below:

```
var SETNODE : SetNode . vars STEPS COMM : Nat .
op moduleName : -> Qid . --- Name of the module with the user's network

crl [step] : STATE => STATE'
    if  EDGES ; SSE := strategy /\
        STATE'     := step([moduleName], STATE, EDGES) .
```

In this rule, the current STATE is updated to STATE' by non-deterministically selecting a set of EDGES from the set of set of edges available according to the strategy. The function `step` below takes as parameters the meta-representation of the user's module defining the network (`[moduleName]`), the current state, and the selected set of edges.

```
var SETAG : SetAgent . var SETOP : SetOpinion . var OP : Opinion .

op step : Module State SetAgent SetOpinion SetEdge -> State .
op step : Module State                    SetEdge -> State .

eq step(M, STATE, EDGES) =
 step(M, STATE, incidents(EDGES), empty, EDGES) .
eq step(M, STATE, empty, SETOP, EDGES) =
   < nodes: (nodes(STATE) / SETOP) ; edges: edges(STATE) >
   in step: (steps(STATE) + 1) comm: (comm(STATE) + | non-self(EDGES) |) .
eq step(M, STATE, (AGENT, SETAG), SETOP, EDGES) =
   step(M, STATE, SETAG, (SETOP, next(M, AGENT, EDGES, STATE)), EDGES) .
```

The function `step` recursively computes the beliefs of the agents incident to EDGES. The updated beliefs are accumulated in the set of opinions SETOP. The opinions of the other agents remain as in STATE (operator /), and the number of steps and the number of communications are updated accordingly. The expression | `non-self(.)` | returns the number of edges that are not self-loops and `nodes(.)` returns the opinions (Γ_o) in a state.

The function `next` computes the outcome of the transition $\langle u : o_u \rangle \rightarrow_A \langle u : o'_u \rangle$ by applying (`metaApply`) the rule `atomic` with the needed substitutions to make this rule executable (and deterministic). Namely, it fixes the opinion to be updated (AGENT and BELIEF), the current STATE, and the set of EDGES to be considered during the update.

```
op next : Module Agent SetEdge State -> Opinion .
ceq  next(M, AGENT, EDGES, STATE) = OP
if  SUBS := 'AGENT:Agent     <- upTerm(AGENT) ;
            'BELIEF:Float    <- upTerm(opinion(AGENT, STATE)) ;
            'STATE:State     <- upTerm(STATE) ;
            'EDGES:SetEdge   <- upTerm(EDGES) /\
   RES? := metaApply(M, upTerm(< AGENT : opinion(AGENT, STATE) >),
           'atomic, SUBS, 0) /\
   OP   := if RES? == failure then error
           else downTerm(getTerm(RES?), error) fi .
```

The opinion function returns the opinion of an agent in a given state.

5 Experimentation

This section shows how Maude and some of its tools can be used to analyze instantiated versions of the rewrite theory $\mathcal{R}$ (see Sect. 4) to better understand the evolution of opinions in networks of agents. Of special interest is checking the (im)possibility of reaching a consensus (i.e., agent's opinions converge to a given value) or stability of the systems, computing the number of steps to reach consensus, computing an optimal strategy to reach consensus, measuring the polarization of the system at each time-step, among others. It is noticed that for De Groot and gossip-like models, there are theoretical results identifying topological conditions that guarantee consensus. In particular, in these models, the agents reach consensus if the graph is strongly connected and aperiodic (i.e., the greatest common divisor of the lengths of its cycles is one) [14].

5.1 Finding Consensus

Let Example-DG be the module/theory extending $\mathcal{R}$ with the following operators and equations:

```
op init : -> Network .                    --- Initial state (as in Fig 1)
eq init = < nodes: ... ; edges: ... > in step: 0 comm: 0 .
eq moduleName = 'Example-DG .             --- Name of the theory

--- Predefined μ for De Groot
eq update(STATE, SETEDGE, AGENT) = deGrootUpdate(STATE, SETEDGE, AGENT) .
eq strategy     = deGroot(edges(init)) . --- De Groot strategy
```

The following command answers the question of whether it is possible to reach a consensus from the initial state. Function consensus(.) checks if all opinions o_i and o_j in a given state satisfy $|o_i - o_j| < \epsilon$, where ϵ is an error bound.

```
Maude> search [1]   init =>* STATE such that consensus(STATE) .

Solution 1 (state 34)
STATE --> < nodes: < 0 : 4.80e-1 >, < 1 : 4.79e-1 >, < 2 : 4.79e-1 >, ...
            edges: <(0,1): 5.99e-1 >, <(0,2): 4.00e-1 >, ... >
            in step: 34 comm: 272
```

The consensus about the given proposition is approximately 0.48 and it is reached in 34 steps. Since in the De Groot model all the 12 edges are considered in each interaction, there is a total of $272 = 34 \times 8$ communications (the interactions on the self-loops are not considered in that counting). Note that an application of rule step in this case is completely deterministic (the strategy considers only one possible outcome, including all the edges of the network).

Let Example-H be as Example-DG, but considering the strategy and update functions for the hybrid model. As explained in Sect. 3.4, the hybrid model exhibits the maximum degree of non-determinism. Using search to check the existence of a reachable state satisfying consensus for the system in Fig. 1 (12 edges) becomes unfeasible: a state may have up to 4095 (nonempty subsets of Γ_i) successor states. Certainly, for this network, a solution must exist due to the above output of the search command and the fact that $\rightarrow_{\text{DeGroot}} \subseteq \rightarrow_{\text{hybrid}}$.

Consider the following rewrite rule and expression in the Maude's strategy language:

```
crl [step'] :   STATE => STATE'
    if STATE' := step([moduleName], STATE, EDGES) [nonexec] .

var STR : SetSetEdge .
strat round : SetSetEdge @ State .
sd round(EDGES ; STR) := (match STATE s.t. consensus(STATE))
                        or-else step'[EDGES <- EDGES] ; round(STR) .
```

Unlike step, rule step' does not use the model strategy to select the set of EDGES that will be used to compute the next state (and hence, it is non-executable). The Maude's strategy round checks whether the current state satisfies consensus and stops. Otherwise, it non-deterministically chooses a set EDGES, applies the rule step' instantiating the set of edges with that particular set, and it is recursively called without EDGES. In other words, round starts with a set of possible interactions and it allows for these interactions to happen only once. This is certainly one of the possible behaviors that can be observed with the hybrid model. Using this strategy, the solutions found by the commands below positively answer the following question for the model in Fig. 1: Can consensus be reached by making some groups of agents (non necessarily disjoint) interact only once? The expression filter>=(n,STR) below returns the sets in STR with cardinality at least n.

```
Maude> dsrew [1] init using round(hybrid(edges)) .
Solution 1
result State: < nodes: < 0 : 0.0 >, ... edges: ... > in step: 8 comm: 13 .

Maude> dsrew [1] init using round(filter>=(6, hybrid(edges))) .
Solution 1
result State: < nodes: < 0 : 1.50e-1 >, ... > in step: 21 comm: 88 .
```

As expected, because of the non-deterministic nature of the hybrid model, the value of consensus (and the number of steps to reach such a state) can heavily depend on the choice of edges at each step. In the first output returned by dsrew in the first command, all the sets considered by round included edges where a

acts as an influencer and the edge $f \to a$ is never selected. This explains the value of the consensus, where the opinion of a was propagated to her neighbors. In the send command, larger groups are chosen to interact, and the edge $f \to a$ is selected in 4 out of the 21 interactions. Hence, a eventually changes her opinion.

5.2 Statistical Analysis

An alternative approach to deal with the inherent state space explosion problem when analyzing $\mathcal{R}$ is to perform statistical model checking. In the following, the tool umaudemc [26] is used for such a purpose. The umaudemc command scheck enables Monte-Carlo simulations of a rewrite theory extended with probabilities; it estimates the value of a quantitative temporal expression written in the query language QuaTEx [1].

Consider the following QuaTEx expression that computes the probability of reaching a consensus before N communications:

```
Prob(N) = if (s.rval("consensus(S)")) then 1.0 else
          if (s.rval("comm(S)") <= N) then # Prob(N) else 0.0 fi fi;
```

The two commands below estimate the probability (output of the tool $\mu = val$) of reaching consensus before 30 (expression E[Prob(30)]) and 20 communications, respectively, in the running example when the gossip-based model is considered. The confidence level of these analyses is 95% and the same probability is assigned to every successor state (--assign uniform).

```
umaudemc  scheck ex-gossip init formula -a 0.05 -d 0.01 --assign uniform
    (μ = 0.587)

umaudemc  scheck ex-gossip init formula -a 0.05  -d 0.01 --assign uniform
    (μ = 0.348)
```

As expected, reducing the maximum number of communications decreases the changes of reaching a consensus state.

The authors in [5] hypothesize that the less dispersed opinion becomes, the easier it will be to reach consensus. In fact, the variance (a standard measure of dispersion) is used as a measure of opinion polarization in social networks [5]. The following commands aim at testing such a hypothesis in the running example when considering the hybrid model:

```
umaudemc  scheck example-H init ... --assign uniform
    (μ = 0.901)

umaudemc  scheck example-H init  ... --assign "term(variance(L,R))"
    (μ = 1.0)

umaudemc  scheck example-H init ... --assign "term(distance(L,R))"
    (μ = 0.987)
```

These commands estimate the probability of reaching consensus before 300 communications (E[Prob(300)]). In the first case, all the successor states are assigned the same probability. In the second, successor states whose set of chosen agents has higher `variance` are assigned higher probabilities. In the third command, successor states whose set of chosen agents are more *polarized*, in the sense that the `distance` between the maximal and the minimal opinions is bigger, are assigned higher probabilities. These results confirm the hypothesis that it is more likely (1.0 vs. 0.9) to reach consensus sooner when communications of agents with more distant opinions is encouraged to reduce dispersion of opinions.

6 Concluding Remarks

This paper presented a unified framework for dynamic opinion models. Such models are tools to analyze the evolution of opinion values, about a given topic, in a network of agents whose opinion may be influenced by other agents. Set relations, which are used for specifying and analyzing concurrent behavior in collections of agents, are the formalism used to unify the modeling of these systems. This framework relies on two mechanisms, namely, an atomic relation that updates the opinion of single agents based on a collection of interactions and a strategy defining the collections of interactions to be considered. The framework is formally specified as a rewrite theory, which is expected to be instantiated for the opinion dynamic model of interest. Three different dynamic opinion models (De Groot, goossip-like, and hybrid) are shown to be instances of this framework. Experiments on these models show that statistical model checking is a promising alternative to tackle the state space explosion problem when analyzing models with a high degree of non-determinism, such is the case of the hybrid model. To the best of the authors' knowledge, this is the first documented effort to make available concurrency theory, techniques, and tools for the specification and analysis of opinion dynamics models and properties such as polarization and consensus.

The ultimate goal of making available computational ideas and approaches for analyzing phenomena in social networks requires (significant) additional work. First, a more in-depth exploration of properties related to these phenomena in social networks is required. This may lead to the proposal of new temporal and probabilistic properties that cannot be handled with current techniques and approaches supporting the opinion dynamic modeling community, but that may be highly supported by the developments in concurrency and computational logics. Second, extensions to the current framework in terms of more general dynamic networks (i.e., the value of influences can change), temporal networks (i.e., nodes and edges can appear and disappear), and the inclusion of several topics/propositions that may share causal relations are in order. Third, more experimental validation is required, ideally with data gathered from real social networks. Fourth, building on the abstract relations proposed here, techniques from concurrency theory become available for the analysis of social systems. It is worth exploring standard concurrency techniques such as bisimulation and testing equivalences to answer questions such as whether two social systems ought to be equivalent and whether there is a social context, represented as a social system, that can tell the difference between two other social systems.

References

1. Agha, G., Meseguer, J., Sen, K.: PMaude: rewrite-based specification language for probabilistic object systems. ENTCS **153**(2), 213–239 (2006)
2. Alvim, M.S., Amorim, B., Knight, S., Quintero, S., Valencia, F.: A formal model for polarization under confirmation bias in social networks. Log. Methods Comput. Sci. **19**(1) (2023)
3. Ballard, A.O., DeTamble, R., Dorsey, S., Heseltine, M., Johnson, M.: Dynamics of polarizing rhetoric in congressional tweets. Legis. Stud. Q. **48**(1), 105–144 (2023)
4. Beaufort, M.: Digital media, political polarization and challenges to democracy. Inf. Commun. Soc. **21**(7), 915–920 (2018)
5. Bramson, A., Grim, P., Singer, D.J., Berger, W.J., Sack, G., Fisher, S., Flocken, C., Holman, B.: Understanding polarization: Meanings, measures, and model evaluation. Philoso. Sci. **84**(1), 115–159 (2017)
6. Center for Strategic and International Studies: The #MilkTeaAlliance in Southeast Asia: Digital revolution and repression in Myanmar and Thailand (2021). https://www.csis.org/blogs/new-perspectives-asia/milkteaalliance-southeast-asia-digital-revolution-and-repression, visited 12-30-2023
7. Clavel, M., Durán, F., Eker, S., Lincoln, P., Martí-Oliet, N., Meseguer, J., Talcott, C.: All About Maude—A High-Performance Logical Framework, LNCS, vol. 4350. Springer, Berlin (2007)
8. Das, A., Gollapudi, S., Munagala, K.: Modeling opinion dynamics in social networks. In: Proceedings of the 7th ACM International Conference on Web Search and Data Mining, pp. 403–412. Association for Computing Machinery, New York, NY, USA (2014)
9. Degroot, M.H.: Reaching a consensus. J. Am. Stat. Assoc. **69**(345), 118–121 (1974)
10. Fagnani, F., Zampieri, S.: Randomized consensus algorithms over large scale networks. In: 2007 Information Theory and Applications Workshop, pp. 150–159 (2007)
11. Fitriani, H.M.: Social media and the fight for political influence in Southeast Asia (2023). https://thediplomat.com/2023/08/social-media-and-the-fight-for-political-influence-in-southeast-asia, visited 12-30-2023
12. Foundation, T.A.: Violent Conflict, Tech Companies, and Social Media in Southeast Asia: Key Dynamics and Responses. The Asia Foundation, San Francisco, USA (2020)
13. Garrett, R.K.: The "echo chamber" distraction: Disinformation campaigns are the problem, not audience fragmentation. J. Appl. Res. Memory Cognition **6** (2017)
14. Golub, B., Sadler, E.: Learning in social networks. Social Science Research Network (SSRN) (2017). http://dx.doi.org/10.2139/ssrn.2919146
15. Gordon-Zolov, T.: Chile's estallido social and the art of protest. Sociologica **17**(1), 41–55 (2023)
16. Gupta, S., Chauhan, V.: Understanding the role of social networking sites in political marketing. Jindal J. Bus. Res. **12**(1), 58–72 (2023)
17. Xia, H., Huili Wang, Z.X.: Opinion dynamics: A multidisciplinary review and perspective on future research. Int. J. Knowl. Syst. Sci. **2**(4), 72–91 (2023)
18. Iversen, T., Soskice, D.: Information, inequality, and mass polarization: Ideology in advanced democracies. Comput. Pol. Stud. **48**(13), 1781–1813 (2015)
19. Kirby, E.: The city getting rich from fake news. BBC News Documentary (2017). https://www.bbc.com/news/magazine-38168281

20. Lynch, M.: After the arab spring: How the media trashed the transitions. J. Democr. **26**(4), 90–99 (2015)
21. Meseguer, J.: Conditional rewriting logic as a unified model of concurrency. Theor. Comput. Sci. **96**(1), 73–155 (1992)
22. Neverov, K., Budko, D.: Social networks and public policy: Place for public dialogue? In: Proceedings of the International Conference IMS-2017, pp. 189–194. Association for Computing Machinery, New York, NY, USA (2017)
23. Olarte, C., Ramírez, C., Rocha, C., Valencia, F.: Opinion dynamic modeling as concurrent set relations in rewriting logic. https://github.com/promueva/maude-opinion-model
24. Rocha, C., Muñoz, C.A.: Synchronous set relations in rewriting logic. Sci. Comput. Program. **92**, 211–228 (2014)
25. Rocha, C., Muñoz, C.A., Dowek, G.: A formal library of set relations and its application to synchronous languages. Theor. Comput. Sci. **412**(37), 4853–4866 (2011)
26. Rubio, R., Martí-Oliet, N., Pita, I., Verdejo, A.: Strategies, model checking and branching-time properties in maude. J. Log. Algebraic Methods Program. **123**, 100700 (2021)
27. Rubio, R., Martí-Oliet, N., Pita, I., Verdejo, A.: Qmaude: Quantitative specification and verification in rewriting logic. In: Chechik, M., Katoen, J., Leucker, M. (eds.) FM 2023. LNCS, vol. 14000, pp. 240–259. Springer, Berlin (2023)
28. Sarma, P., Hazarika, T.: Social media and election campaigns: an analysis of the usage of twitter during the 2021 Assam assembly elections. Int. J. Soc. Sci. Res. Rev. **6**(2), 96–117 (2023)
29. Suresh, V.P., Nogara, G., Cardoso, F., Cresci, S., Giordano, S., Luceri, L.: Tracking fringe and coordinated activity on twitter leading up to the us capitol attack (2023)
30. Wikipedia Foundation: 2021 Colombian protests. https://en.wikipedia.org/wiki/2021_Colombian_protests, visited 12-30-2023

Timed Strategies for Real-Time Rewrite Theories

Carlos Olarte[1](✉) and Peter Csaba Ölveczky[2]

[1] LIPN, CNRS UMR 7030, Université Sorbonne Paris Nord, Villetaneuse, France
olarte@lipn.univ-paris13.fr
[2] Department of Informatics, University of Oslo, Oslo, Norway

Abstract. We propose a language for conveniently defining execution strategies for real-time rewrite theories, and provide Maude-strategy-implemented versions of most Real-Time Maude analysis methods, albeit with user-defined discrete and timed strategies. We also identify a new time sampling strategy that should provide efficient and exhaustive analysis for many distributed real-time systems. We exemplify our language and its analyses on a simple round trip time protocol, and compare the performance of standard Maude search with our strategy-implemented reachability analyses on the CASH scheduling algorithm benchmark.

1 Introduction

Real-time systems can naturally be defined in rewriting logic [22] as *real-time rewrite theories* [29]. In such theories, actions that can be assumed to take zero time are modeled by ordinary (also called *instantaneous*) rewrite rules, and time advance is modeled by labeled "tick" rewrite rules of the form $[l] : \{t_1\} \longrightarrow \{t_2\}$ **in time** τ **if** *cond*, where the *whole* system state has the form $\{t\}$.

Real-time rewrite theories inherit the expressiveness and modeling convenience of rewriting logic, and allow us to model a wide range of distributed real-time systems—with different communication forms, user-defined data types, dynamic object creation and deletion, and so on—in an object-oriented style.

The specification and analysis of real-time rewrite theories is supported by the Real-Time Maude language and tool [26,30,31], which is implemented in Maude as an extension of Full Maude [12]. For dense time, the tick rules typically have the form `crl [tick]  : {`t`} => {`u`} in time T if T <=` $f(t)$, where T is a variable of sort `Time` not appearing in the term t [30].

Real-Time Maude provides *explicit-state* analysis methods, where the above tick rules are executed according to a *time sampling strategy*, where the variable T in the rule is always instantiated to either a *user-selected value* (such as 1) or the *maximal possible time increase* $f(t)$. Real-Time Maude supports unbounded and time-bounded reachability analysis, LTL and timed CTL model checking, and other time-specific analyses. All analyses are performed with the selected time sampling strategy, and may not cover all possible system behaviors.

Real-Time Maude has been used to discover subtle but significant bugs in a number of sophisticated systems beyond the scope of decidable formalisms like timed automata, including: a 50-page active network protocol [32] (which required advanced functions and detailed modeling of communication), a wireless sensor network algorithm [33] (involving functions on coverage areas), a mobile ad-hoc network protocol [20] (the fault was due to a subtle interplay between node movements and communication delays), scheduling algorithms with reuse of unused budgets [28] (which required unbounded queues), a traffic intersection system from the Ptolemy II library [5,18] (which required defining the semantics of Ptolemy II discrete-event models), cloud-based transaction systems [8,16], and an error in cars, where Real-Time Maude time sampling was key.[1]

Real-Time Maude's expressiveness and generality have also allowed it to provide formal semantics and formal analysis capabilities to (subsets of) modeling languages such as AADL [27], Ptolemy II DE models [5], Timed Rebeca [38], and a DoCoMo Labs handset language [1].

In this paper we use Maude's strategy language [15] to define useful strategies for real-time rewrite theories. This work is motivated by the following issues:

1. Real-Time Maude analyses apply one of the above two time sampling strategies to *all* applications of tick rules. However, as the following example shows, more sophisticated time sampling strategies are often desired:
   Consider a system computing the *round trip time* (RTT) between two nodes every five seconds. A time sampling strategy that visits each time unit covers all possible behaviors in discrete time domains, but visits each time point even when the round trip time in that round already has been found. On the other hand, increasing time maximally in each tick step only takes into account behaviors where each message has been delayed as much as possible. In this simple but prototypical example, the ideal time sampling strategy advances time by one time unit when there is a message (which could arrive at "any" time) in the state, and increases time maximally otherwise (when we are just idling until the next iteration of the protocol). Such time sampling would cover all possible behaviors, yet would not stop time unnecessarily.
2. The user may also want to define *non-time-sampling* execution strategies, such as *eagerness* of all/some actions, or giving priority to some actions.
3. Maude's strategy language has an efficient implementation, using multi-threading and the option of depth-first search analyses, which provides better performance than standard Maude search in some cases (see Sect. 5.6).
4. Real-Time Maude is implemented as an extension of Full Maude to support object-oriented specification. Since Maude 3.3 supports object-oriented specification, and since Full Maude will no longer be maintained, we are working on developing the next version of Real-Time Maude as a Maude implementation that does not extend Full Maude. In this context, doing as much as possible as easily as possible using available Maude features is needed.

In this paper we therefore show how most Real-Time Maude analysis methods can be performed by rewriting with strategies directly in Maude (Sect. 4).

---

[1] Hitoshi Ohsaki, personal communication, 2007.

However, even those analysis methods needed somewhat hard-to-understand strategy expressions. This begs the question how the casual Maude user can analyze her system with more complex discrete and time sampling strategies. Furthermore, since Real-Time Maude provides formal analyses for many modeling languages, we need to allow the non-expert Maude user to define her strategies. For example, in [3,4] we claim the ability to analyze the system with user-defined execution strategies as a selling point of our Maude framework for parametric timed automata and time Petri nets. However, this selling point becomes moot if the timed automaton/Petri net expert cannot define her strategies.

To address this issue, in Sect. 5 we define and implement what we hope is an intuitive and fairly powerful timed strategy language for real-time rewrite theories. This language should make it easy for the casual user to define a wide range of useful discrete strategies as well as advanced state- and even history-dependent time sampling strategies in a modular way. In Sect. 5.6 we compare the performance of standard Maude search with our strategy-implemented analysis methods on a sophisticated scheduling algorithm [28].

We discuss related work in Sect. 6, and give some concluding remarks in Sect. 7. The strategy language, with Maude models and execution commands, are available at [24]. Due to space restrictions, we only present parts of our contribution in this paper, and refer to our report [25] for more detail.

2 Preliminaries

Rewriting Logic and Maude. Maude [13] is a rewriting-logic-based executable formal specification language and high-performance analysis tool for distributed systems. A Maude module specifies a *rewrite theory* (Σ, E, R), where:

- Σ is an algebraic *signature*; i.e., a set of *sorts*, *subsorts*, and *function symbols*.
- (Σ, E) is a *membership equational logic* [23] theory, with E a set of possibly conditional equations and membership axioms.
- R is a collection of *labeled conditional rewrite rules* $[l] : t \longrightarrow t'$ if $cond$, specifying the system's local transitions.

A function f is declared op f : $s_1 \ldots s_n$ -> s .Equations and rewrite rules are introduced with, respectively, keywords eq , or ceq for conditional equations, and rl and crl . Mathematical variables are declared with the keywords var and vars , or can have the form *var*:*sort* and be introduced on the fly. class C | att_1 : s_1, ... , att_n : s_n declares a *class* C of objects with attributes att_1 to att_n of sorts s_1 to s_n. An *object instance* of class C is represented as a term < O : C | att_1 : $val_1, \ldots, att_n$: val_n >, where O is the object's *identifier*, and where val_1 to val_n are the values of the attributes att_1 to att_n. A *message* is a term of sort Msg. A system state is modeled as a term of sort Configuration, and has the structure of a *multiset* made up of objects and messages.

Formal Analysis in Maude. Maude provides a number of analysis methods, including rewriting for simulation purposes, reachability analysis, and linear temporal logic (LTL) model checking. The command red *expr* reduces the expression *expr* to its *E*-normal form. Given a state pattern *pattern* and an (optional) condition *cond*, Maude's command search *init* =>* *pattern* [such that *cond*] searches the reachable state space from *init* for all (or optionally a given number of) states that match *pattern* such that *cond* holds.

Strategies. Maude provides a language for defining strategies to control and restrict rewriting. A strategy may not make rewriting deterministic, and hence multiple behaviors allowed by the strategy must be explored. The command srew *t* using *str* rewrites the term *t* according to the strategy *str*, and returns a set of terms, possibly bounded by the number of desired solutions. srew explores multiple paths in parallel, and ensures that solutions will eventually be found. dsrew *t* using *str* explores the behaviors allowed by *str* in a depth-first way.

Basic rewrite strategies include $l[\sigma]$ (apply rule labeled l once with the optional substitution σ), all (apply any of the rules, except those marked nonexec, once), idle (identity), fail (empty set), and match *pattern* s.t. *cond*, which checks whether the current term matches the *pattern* subject to the constraint *cond*. Compound strategies can be defined using concatenation (α ; β), disjunction (α | β, whose result is the union of the results of α and β), iteration ($\alpha*$), α or-else β (execute α, and β if α fails), try(α) (applies α if it does not fail), normalization α! (execute α until it cannot be further applied), matchrew $p(x_1,\ldots,x_n)$ s.t. *cond* by x_1 using α_1, $\ldots$, x_n using α_n (if the term matches the pattern $p(x_1,\ldots,x_n)$, then, for each match σ, rewrite each substitution instance $x_i\sigma$ in the term according to the strategy α_i), and so on [12].

Metaprogramming. Maude supports metaprogramming in the sense that a Maude specification M can be represented as a *term* $\overline{M}$ (of sort Module), and a term t in M can be (meta-)represented as a term $\overline{t}$ of sort Term. Maude's META-LEVEL module contains a number of useful meta-level versions of key Maude functionality, including metaSrewrite (srew and dsrew at the meta-level).

Real-Time Rewrite Theories and Real-Time Maude. Real-time systems can be defined in rewriting logic as real-time rewrite theories [29], which are parametric in the (discrete or dense) time domain. In such theories, ordinary rewrite rules model *instantaneous* change, and time advance is modeled by "tick" rewrite rules crl [*tick*] {*t*} => {*u*} in time τ if *cond* , where τ is a term of sort Time, t and u are terms of sort System, the *entire* state has the form {*s*}, and {_} does not occur in *s*; this ensures that time advances uniformly in the whole system.

Real-Time Maude [26,30,31] supports the modeling and analysis of real-time rewrite theories. Most Real-Time Maude specifications have tick rules

crl [*tick*] : {$t(\overline{x})$} => {$u(\overline{x},y)$} in time y if y <= $f(t(\overline{x}))$ /\ *cond* [nonexec] .

where y is a variable that does not appear in t and is not instantiated in *cond*, making the rule non-executable (nonexec) as it stands.

Real-Time Maude therefore offers the user the possibility of choosing between the following *time sampling strategies* for executing such *time-nondeterministic* tick rewrite rules:

- *deterministic time sampling* with a user-given time value $\delta > 0$; and
- *maximal time sampling*, with a user-given "default" time value $\delta > 0$.

Using *deterministic* time sampling, the variable y in the above tick rule is instantiated by the selected time value δ in each application of a time-nondeterministic tick rule; the tick rule cannot be applied to a state $\{s\}$ if $f(s)$ is smaller than δ.

Using *maximal* time sampling, the variable y is instantiated to advance time as much as possible, namely, by $f(s) > 0$ in state $\{s\}$, unless $f(s)$ is the infinity value INF, in which case y is instantiated with the "default" time value δ instead.

Real-Time Maude provides the following analysis methods, where the *same* selected time sampling strategy is applied in *all* tick rule applications [30]:

- *Rewriting* up to time Λ.
- *Time-bounded and untimed search* for states matching a pattern $p(\overline{x})$, such that an optional condition $cond(\overline{x})$ holds, that are reachable from the initial state (possibly within a given time interval $[l, u]$).
- *Time-bounded and unbounded LTL model checking* check whether each behavior from *init*, up to a given time bound in the time-bounded case, satisfies an (untimed) linear temporal logic (LTL) formula.
- *Timed CTL model checking* checks whether each behavior, possibly up to a user-given time bound, satisfies a given *timed* CTL formula [18].
- *Find latest* finds the longest time it takes to reach a desired state.
- *Find earliest* finds the shortest time needed to find the desired state.

In time-bounded Real-Time Maude analyses, internally the state also contains the "system clock" denoting the time it takes to reach the corresponding state.

3 The Targeted Real-Time Rewrite Theories

This section presents assumptions about the real-time rewrite theories we consider in this paper. Most large Real-Time Maude applications belong to this class of real-time rewrite theories, or can easily be modified to do so (e.g., by renaming rule labels and variables). We then present our running example as one such model of a prototypical real-time system: a simple protocol for computing the *round trip times* between pairs of senders and receivers in a network.

Assumptions. We specify our real-time rewrite theories directly in Maude by extending the following "timed prelude," which defines the sorts of our states:

```
fmod TIMED-PRELUDE is including TIME .
   sorts System GlobalSystem ClockedSystem .
   subsort GlobalSystem < ClockedSystem .

   op {_} : System                        -> GlobalSystem  [ctor] .
   op _in time_ : GlobalSystem Time -> ClockedSystem [ctor] .
```

```
    var CLS : ClockedSystem .      vars T T' : Time .
    eq (CLS in time T) in time T' = CLS in time (T plus T') .
endfm
```

We assume a sort Time for the time values, a supersort TimeInf adding an infinity element INF to those values, and assume that each tick rule has the form

```
var T : Time .
crl [tick] : {t} => {u} in time T  if T <= mte(t) /\ cond [nonexec] .
```

or the form rl [tick] : {t} => {u} in time T [nonexec], where the symbols in italics are placeholders for terms and conditions. In particular, we assume that the unknown time advance is represented by the specific variable T (not appearing in t nor in cond), that all tick rules are labeled tick, that no non-tick rule is labeled tick, and that the maximal time elapse is given by the (user-defined) function mte, which returns a time value or INF.

Running Example: Finding Round Trip Times. The following object-oriented Maude model specifies a simple protocol for computing the *round trip time* (RTT) between pairs of Senders and Receivers every 5 seconds. The delay of a message can be any value between a lower and an upper bound. This small example contains many features of larger real-time distributed protocols: clocks, timers, and messages with nondeterministic delays.

```
omod RTT is
  including TIMED-PRELUDE .   protecting NAT-TIME-DOMAIN-WITH-INF .

  var M     : Msg .   var TI   : TimeInf .      vars T T2 T3 : Time .
  vars R S  : Oid .   vars C1 C2 STATE : Configuration .

  sort DlyMsg .  subsorts Msg < DlyMsg < Configuration < System .
  op dly : Msg Time Time -> DlyMsg [ctor] .  --- upper and lower bounds

  rl [deliver] : dly(M, 0, TI) => M .  --- deliver ripe message any time

  msgs rttReq_from_to_ rttResp_from_to_ : Time Oid Oid -> Msg .

  class Sender | clock : Time, timer : Time, lowerDly : Time, period : Time,
                 upperDly : TimeInf, rtt : TimeInf, receiver : Oid .

  class Receiver | lowerDly : Time, upperDly : TimeInf .

  rl [send] : < S : Sender | clock : T, timer : 0, period : T2,
                             lowerDly : T3, upperDly : TI, receiver : R >
       =>   < S : Sender | timer : T2 > dly(rttReq T from S to R, T3, TI) .

  rl [respond] :
       (rttReq T from S to R) < R : Receiver | lowerDly : T3, upperDly : TI >
       =>   < R : Receiver | > dly(rttResp T from R to S, T3, TI) .

  rl [recordRTT] : (rttResp T from R to S) < S : Sender | clock : T2 >
               =>   < S : Sender | rtt : T2 monus T > .

  crl [tick] : {STATE}  => {timeEffect(STATE, T)} in time T
               if T <= mte(STATE) [nonexec] .
```

```
   op mte : Configuration -> TimeInf [frozen] .
   eq mte(< S : Sender | timer : TI >) = TI .
   eq mte(dly(M, T, TI)) = TI .
   eq mte(M) = 0 .        --- ripe message must be read immediately
   ...  --- see our report for the other equations for mte and timeEffect
   op timeEffect : Configuration Time -> Configuration .
   eq timeEffect(< S : Sender | clock : T, timer : TI >, T2)
    = < S : Sender | clock : T + T2, timer : TI monus T2 > .
   eq timeEffect(dly(M, T, TI), T2) = dly(M, T monus T2, TI monus T2) .

   ops snd rcv : -> Oid [ctor] .       op init : -> ClockedSystem .
   eq init
    = {< snd : Sender | clock : 0, timer : 0, period : 5000, lowerDly : 5,
                        upperDly : 20, rtt : INF, receiver : rcv >
       < rcv : Receiver | lowerDly : 7, upperDly : 30 >} in time 0 .
endom
```

A `Sender` object has the following attributes: `clock` denotes its "local clock;" `timer` denotes the time until its next round begins; `lowerDelay` and `upperDelay` are bounds on the delays of messages from the sender; `rtt` stores the latest round trip time value; `period` denotes its period; and `receiver` denotes its receiver. A `Receiver` object has attributes for the delays of messages *from* the receiver.

A "delayed" message $\mathtt{dly}(m, t, t')$ denotes a message m whose *remaining* delay is in the interval $[t, t']$. The rule `deliver` removes the `dly` wrapper, thereby making the message "ripe," whenever the lowest remaining delay has reached 0.

When a `Sender`'s `timer` expires (i.e., becomes 0), a new round of the RTT-finding protocol starts (rule `send`). The `Sender` sends an `rttReq` message with its current clock value `T` to the `Receiver`, and its timer is reset. When a `Receiver` receives such a request, it replies with an `rttResp` message (rule `respond`). When a `Sender` receives this response, with its original timestamp `T`, it can easily compute and store the (latest) round trip time (rule `recordRTT`).

The tick rule in this system, which could have many `Sender`/`Receiver` pairs, is the usual one for object-oriented Real-Time Maude specifications [30]. `mte` ensures that time cannot pass beyond the time when a message *must* be delivered, that time cannot pass when there is a "ripe" (un-delayed) message in the state, and that time cannot pass beyond the expiration time of any `timer`. The function `timeEffect` reduces the remaining bounds of all message delays and timer values, and increases the `clock` values, according to the elapsed time.

`init` defines an initial state with one `Sender` and one `Receiver`.

4 Analysis Using Maude's Strategy Language Directly

Our report [25] explains how we can perform "Real-Time Maude-style" time-sampling-strategy-based (time-bounded, unbounded, and "clock-less" unbounded) reachability analysis, as well as time-bounded simulation, using Maude's strategy language. Due to space restrictions, we just show timed-bounded reachability with maximal time sampling, and refer to [25] for the other cases.

Example 1. We search for two states where the RTT value 50 can be found in the time interval [5000, 10000] using maximal time sampling:[2]

```
Maude> srew [2] init using (all |
   (matchrew CS:ClockedSystem
       such that {STATE} in time T2 := CS:ClockedSystem /\ mte(STATE) =/= 0
           /\ T2 + (if mte(STATE) == INF then 4 else mte(STATE) fi) <= 10000
       by CS:ClockedSystem using
           tick[T <- if mte(STATE) == INF then 4 else mte(STATE) fi])) *
  ; (match  {< snd : Sender | rtt : 50, ATTS:AttributeSet >
             C:Configuration} in time T3:Time s.t. T3:Time >= 5000) .

Solution 1
result ClockedSystem:
 {< snd : Sender | rtt : 50, ... >   < rcv : Receiver | ... >}  in time 5000

Solution 2    ...
```

The above strategy repeatedly (*) applies any rule, followed by (;) checking whether the results `match` the pattern where the sender's `rtt` attribute is 50 and the "system clock" is greater than or equal to 5000. Regarding the application of any rewrite rule, it either applies any executable rewrite rule (`all`) or (`|`) a tick rule. If it is a `tick` rule, the variable `T` denoting the time increase is instantiated to `mte(STATE)`, for the given `STATE` obtained using `matchrew`, unless `mte(INF)` equals `INF`, in which case `T` is set to 4. The tick rule is not applied if `mte(STATE)` is 0, or if the system clock would go beyond the upper time limit 10000.

The same analysis can be performed with deterministic time sampling, say, with increment 1, by replacing `if mte(STATE) == INF then 4 else mte(STATE) fi` in the above expression with `1`. We obtain unbounded reachability analysis by removing the two tests for time limits above. Finally, for unbounded analyses, we add the following rule, which removes the "system clock", and always apply this rule right after applying a tick rule:

```
rl [removeClock] : {STATE} in time T => {STATE} [nonexec] .
```

We mark this rule `nonexec` to allow us to control when it can be applied (e.g., it will not be applied by the strategy `all`). See [25] for details and examples.

5 A Strategy Language for Real-Time Rewrite Theories

Section 4 shows that even simple reachability analysis needs somewhat hard-to-understand strategy expressions. How can the non-Maude-expert analyze her system with more complex strategies? To address this question, this section defines what we hope is a powerful yet intuitive timed strategy language for real-time rewrite theories that supports: **(1)** *separate* definitions of strategies for *discrete behaviors*, including the interplay between discrete actions and time advance, and *timed strategies*; **(2)** *state-dependent time sampling* strategies and *conditional* discrete strategies; **(3)** *history-dependent* strategies; and **(4)** intuitive syntax for "Real-Time Maude commands" with user-defined strategies.

---

[2] Parts of Maude code and Maude output will be replaced by '...'.

Section 5.1 discusses execution strategies for both real-time systems in general and real-time rewrite theories, and Sect. 5.2 introduces our strategy language. Section 5.3 defines a formal semantics for our strategy language by translating its expressions into expressions in Maude's strategy language. Sections 5.4 and 5.5 show how most Real-Time Maude analysis methods can be performed using our strategy language. Section 5.6 compares the performance of our analysis commands with standard Maude search on the CASH scheduling algorithm. The executable Maude specification is available at [24].

5.1 Strategies for Real-Time Systems

Interesting execution strategies of timed systems in general include:

1. *Eagerness* of certain (or all) actions: time should not advance when such actions can be taken.
2. Advance time (or "idle") by $f(s)$ in all states s belonging to a set of states S, advance time by $g(s')$ in all states $s' \in S'$, and so on.
3. Do not perform action a more/less than x times.
4. Always execute action a_i before action a_j when both are enabled.

These examples indicate that we can consider three "types" of strategies: (i) strategies on the "discrete behaviors" (such as items 3 and 4 above); (ii) strategies on how much to advance time (item 2); and (iii) combining these (item 1). This means that the user may want to specify a strategy restricting the discrete behaviors of a system, as well as a strategy for how to advance time. Therefore, we must be able to *compose* any *discrete strategy* with any *timed strategy*.

In Real-Time Maude, the selected time sampling strategy is used in all tick rule applications. However, with maximal time sampling we may miss too many behaviors, whereas with deterministic time sampling we may cover all possible behaviors (for discrete time), but at the cost of "visiting" each time point, even when the system is just "idling."

An efficient time sampling strategy that covers all (interesting) behaviors for discrete time is the following instance of item (2) above:

- increment time by 1 when an action *could* happen (in the next time instant);
- increment time maximally otherwise.

In the RTT system, we should increment time by 1 when there is a delayed message in the state[3], and maximally when there is no message in the state (and the system is just idling until the next period begins). This suggests an efficient time sampling strategy for a large class of distributed real-time systems [19].

5.2 Our Timed Strategy Language

A strategy $\langle \mu, \tau \rangle$ (of sort UStrat) in our timed strategy language consists of a user-defined *discrete strategy* μ (of sort UDStrat), controlling the way instantaneous rules are applied and their interaction with time passage, and a *timed strategy* τ (of sort UTStrat) defining a time sampling strategy:

---

[3] A further optimization would advance time to when the least remaining delay is 0.

```
sorts UStrat UTStrat UDStrat .
op <_,_> : UDStrat UTStrat -> UStrat .
```

The discrete strategy μ controls whether some (and if so, which) action/instantaneous rule must be applied in the current state, or whether some tick rule must be applied. The timed strategy τ defines exactly how each "tick rule application" (i.e., each delay step) in the discrete strategy μ is applied.

We extend the global state of the system with a map that stores information about the execution history. This allows us to define *history-dependent strategies*, i.e., strategies that depend on the current and the previously visited states:

```
sort StrState .
pr MAP{K, V} * (sort Entry{K, V} to Entry, sort Map{K, V} to Map) .
op _|_ : ClockedSystem Map -> StrState .
```

The sorts K and V for the keys and their values are user-defined.

Discrete Strategies. Discrete strategies are defined using the following syntax.
```
sorts Interval SCond .   --- Intervals and conditions
op  [_,_]                    : Time Time              -> Interval .
op  matches_s.t._            : ClockedSystem Bool     -> SCond .
op  matches_s.t._            : StrState Bool          -> SCond .
op  matches_s.t._            : Map Bool               -> SCond .
op  matches                  : ClockedSystem          -> SCond .
op  in_                      : Interval               -> SCond .
ops after before after= before= : Time                -> SCond .
ops _/\_ _\/_                : SCond SCond            -> SCond .
op  not_                     : SCond                  -> SCond .
--- User-defined strategies
op  apply_                   : Qid                    -> UDStrat .
ops apply[_] eager[_]        : QidList                -> UDStrat .
ops action delay eager stop skip :                    -> UDStrat .
ops _;_  _or_  _or-else_     : UDStrat UDStrat        -> UDStrat .
op  if_then_else_            : SCond UDStrat UDStrat  -> UDStrat .
op  get_and set_             : Map Map                -> UDStrat .
```

Terms of sort SCond define conditions in some of the strategies. The condition matches P s.t. C, where P is a pattern and C is an (optional) Boolean condition, checks whether the current state matches P so that C holds in the state. The pattern P can be a ClockedSystem, or a StrState (a clocked system extended with a Map). Other basic conditions include checking whether the current value t of the global clock satisfies: $t \in [a,b]$ (in [a , b]), $t > t'$ (after t'), $t \geq t'$ (after= t'), $t < t'$ (before t'), and $t \leq t'$ (before= t'). Larger conditions can be constructed using conjunction, disjunction, and negation.

User-defined discrete strategies are: apply ℓ applies the instantaneous rule with label ℓ once; apply [$\mathcal{L}$] applies *once* the first rule in the list of labels $\mathcal{L}$ that succeeds in the current state (i.e., $\mathcal{L}$ defines a *priority* on the next rule to be applied); action applies *any* instantaneous rule once; delay applies a tick rule once; eager applies the instantaneous rules as much as possible, followed by *one* "delay" when it is possible; eager [$\mathcal{L}$] applies as much as possible the rules in the list $\mathcal{L}$ followed by one "delay"; μ ; μ' is the sequential composition of two strategies; μ or μ' returns the union of the results obtained from the strategies μ and μ'; μ or-else μ' applies μ, but applies the strategy μ' if μ

fails; if ϕ then μ else μ' is the conditional strategy; stop is the strategy that always fails; skip leaves the current state unchanged; and get M and set M' uses the pattern M to retrieve (part of) the map storing information about the execution of the strategy and updates it according to M'.

Time Sampling Strategies. Time sampling strategies are defined as follows:

```
sorts CTStrat LCTStrat .      subsort CTStrat < LCTStrat .
op fixed-time_               : Time              -> UTStrat .
op max-time with default_    : Time              -> UTStrat .
op when_do_                  : SCond UTStrat     -> CTStrat .
op switch_otherwise_         : LCTStrat UTStrat  -> UTStrat .
```

fixed-time t advances the time by time t in each application of a tick rule (when advancing time by that amount is possible). max-time with default t advances time in a tick rule application by the maximal time t' possible *for that tick rule*, and advances time by t if t' is INF. The conditional time sampling strategy switch *cases* otherwise τ, where *cases* is a list of choices of the form when ϕ_j do τ_j, executes the first strategy τ_i whose guard ϕ_i holds in the current state; the strategy τ is applied if none of the guards hold.

Example 2. The basic strategy < delay or action , τ >, for any timed strategy τ, applies *any* enabled rule once, and < eager ; τ > prioritizes the application of instantaneous rules over tick rules. We can give preference to the rules send and respond, then to the other actions, and finally to the tick rule:

    < (apply ['send 'respond] or-else action or-else tick, τ >

Regarding the time sampling strategy for our RTT example, for any discrete strategy μ, a good choice for this system is: if there is a *delayed* message in the state, increase time by 1, otherwise increase time maximally. This *state-dependent* time sampling strategy can be defined as follows:

```
< μ, switch when matches ({CONF dly(M,T1,T2)} in time R) do fixed-time 1
      otherwise max-time with default 1 >
```

We could save bandwidth by not performing the RTT-finding procedure in *each* period. We therefore add the following rule to the module RTT; this rule allows a Sender to skip a round of the protocol by just resetting the timer when it expires (instead of also sending an rttReq message):

```
var S : Oid .     vars T T2 : Time .
rl [skipRound] : < S : Sender | timer : 0, period : T2 >
             => < S : Sender | timer : T2 > .
```

When its timer expires, a sender nondeterministically chooses between executing a round of the protocol (rule send) or skipping one round (rule skipRound).

A sensible strategy is to skip some rounds but never skip more than two rounds in a row. To define this *state- and history-dependent* strategy, we use a counter labeled with 'c to avoid skipping "more than two rounds":

```
< delay or
  if matches {< S : Sender | timer : 0, ATTS >} in time T  --- State dep.
  then if (matches ('C |-> N) s.t. N <= 1) --- History dependent
       then apply 'skipRound ;
            (get ('C |-> N) and set ('C |-> N + 1))  --- Skip and increment
       else apply 'send ;
            (get ('C |-> N) and set ('C |-> 0))  --- Send and reset
  else action ) , τ >
```

5.3 Semantics of Our Timed Strategy Language

This section shows how expressions in our timed strategy language can be translated into expressions in Maude's strategy language. The denotational and operational semantics of the latter [15] therefore formally describes the execution of real-time rewrite theories controlled by a timed strategy $\langle \mu, \tau \rangle$.

We define a map $[\![-]\!]$ from terms of sort UStrat to terms of sort Strategy, the sort in Maude's prelude used to meta-represent strategies.

Definition 1. *(Semantics)* The interpretation of conditions ($[\![-]\!]_b$), time sampling strategies ($[\![-]\!]_t$), and real-time strategies ($[\![-]\!]$), as terms of sort Strategy, is given in Fig. 1. These definitions use the following variables, and require the new operator and rule below:

```
vars M M' M'' : Map .  var CS : ClockedSystem .  var SS : StrState .
var B : Bool .  vars Te Te' : Term .  vars T T' T1 T2 : Time .
var C : SCond .  var LC : LCTStrat .  var S : System .
op matching_s.t._ : Term Term -> SCond .
rl [updateMap] : CS | (M, M') => CS | (M, M'') [nonexec] .
```

In Fig. 1, $\bar{t}$ denotes the meta-representation of a term t (upTerm(t)). For instance, the second case in Fig. 1a must be read, and specified in Maude as:
$[\![\text{matching CS s.t. B}]\!]_b = [\![\text{matching '\_|\_[$\overline{CS}$, 'M:Map] s.t. }\overline{B}]\!]_b$

```
eq enc(matching CS s.t. B) =
   enc(matching '_|_[upTerm(CS), 'M:Map] s.t. upTerm(B)) .
```

The Maude strategy $[\![\phi]\!]_b$ fails when the condition ϕ does not hold, and succeeds (without modifying the current state) otherwise. matches expressions are reduced until their parameters are of sort Term. Then, Maude's strategy match is used to check whether the current state matches the pattern and satisfies the given condition (otherwise, match fails).

The Maude strategy $[\![\tau]\!]_t$ applies the tick rule by instantiating the variable T with the needed substitution according to τ. In max-time, Maude's strategy matchrew is used to do pattern matching and bind the variable S with the current configuration. Hence, the call mte(S) determines the next tick value. The definition of switch uses the conditional Maude strategy $\alpha ? \beta : \gamma$ to choose the right time sampling strategy τ_i.

The Maude strategy $[\![\langle \mu, \tau \rangle]\!]$ fails when $\mu = $ stop and does nothing if $\mu = $ skip. If $\mu = $ apply Q, the rule with label Q is applied, without any substitution ([none]) and with the {empty} list of strategies (since no particular

⟦matches SS s.t. B⟧$_b$ = ⟦matches $\overline{\text{SS}}$ s.t. $\overline{\text{B}}$⟧$_b$
⟦matches CS s.t. B⟧$_b$ = ⟦matches $\overline{\text{CS | M}}$ s.t. $\overline{\text{B}}$⟧$_b$
⟦matches M s.t. B⟧$_b$ = ⟦matches $\overline{\text{CS | (M , M')}}$ s.t. $\overline{\text{B}}$⟧$_b$
⟦matches Te s.t. Te'⟧$_b$ = match Te s.t. Te'
⟦after(T)⟧$_b$ = ⟦match { CS } in time T' s.t. T > T'⟧$_b$
⟦$\phi_1 \land \phi_2$⟧$_b$ = ⟦ϕ_1⟧$_b$; ⟦ϕ_2⟧$_b$ ⟦$\phi_1 \lor \phi_2$⟧$_b$ = ⟦ϕ_1⟧$_b$ or-else ⟦ϕ_2⟧$_b$
⟦not ϕ⟧$_b$ = not ⟦ϕ⟧$_b$

(a) Conditions. Definitions for **before**, **in**, etc., are similar and are omitted.

⟦fixed-time T1⟧$_t$ = 'tick [$\overline{\text{T}}$ ← $\overline{\text{T1}}$] { empty }
⟦max-time with default T1⟧$_t$ = matchrew $\overline{\text{SS}}$ s.t. $\overline{\text{{ S } in time T2 | M}}$:= $\overline{\text{SS}}$
   by $\overline{\text{SS}}$ using 'tick [$\overline{\text{T}}$ ← $\overline{\text{if INF == mte(S) then T1 else mte(S) fi}}$] { empty }
⟦switch (when C do τ) LC otherwise τ'⟧$_t$ = ⟦C⟧$_b$? ⟦τ⟧$_t$: ⟦switch LC otherwise τ'⟧$_t$
⟦switch (when C do τ) otherwise τ'⟧$_t$ = ⟦C⟧$_b$? ⟦τ⟧$_t$: ⟦τ'⟧$_b$

(b) Timed strategies.

⟦⟨stop, τ⟩⟧ = fail   ⟦⟨skip, τ⟩⟧ = idle   ⟦⟨apply Q, τ⟩⟧ = Q [none] {empty}
⟦⟨action, τ⟩⟧ = all   ⟦⟨delay, τ⟩⟧ = ⟦τ⟧$_t$ ⟦⟨eager, τ⟩⟧ = all ! ; try(⟦τ⟧$_t$)
⟦⟨apply [nil], τ⟩⟧ = fail   ⟦⟨apply [Q LQ], τ⟩⟧ = apply Q or-else ⟦apply [LQ]⟧
⟦⟨eager [L], τ⟩⟧ = ⟦⟨apply [L], τ⟩⟧ ! : try(⟦τ⟧$_t$)
⟦⟨μ ; μ', τ⟩⟧ = ⟦⟨μ, τ⟩⟧ ; ⟦⟨μ', τ⟩⟧   ⟦⟨μ or μ', τ⟩⟧ = ⟦⟨μ, τ⟩⟧ | ⟦⟨μ', τ⟩⟧
⟦⟨μ or-else μ', τ⟩⟧ = ⟦⟨μ, τ⟩⟧ or-else ⟦⟨μ', τ⟩⟧
⟦⟨if C then μ else μ', τ⟩⟧ = ⟦C⟧$_b$? ⟦⟨μ, τ⟩⟧ : ⟦⟨μ', τ⟩⟧
⟦⟨get M' and set M'', τ⟩⟧ = matchrew $\overline{\text{SS}}$ s.t. $\overline{\text{{ S } in time T1 | (M, M')}}$:= $\overline{\text{SS}}$
   by $\overline{\text{SS}}$ using 'updateMap [$\overline{\text{M}}$ ← $\overline{\text{M}}$; $\overline{\text{M'}}$ ← $\overline{\text{M'}}$; $\overline{\text{M''}}$ ← $\overline{\text{M''}}$] { empty }

(c) Discrete and real-time strategies.

⟦⟨check ϕ, τ⟩⟧ = ⟦ϕ⟧$_b$ ⟦⟨until ϕ do μ, τ⟩⟧ = (⟦ϕ⟧$_b$? fail : ⟦⟨μ, τ⟩⟧)!
⟦⟨repeat μ, τ⟩⟧ = ⟦⟨μ, τ⟩⟧) *   ⟦⟨0 steps with μ, τ⟩⟧ = idle
⟦⟨s(N) steps with μ, τ⟩⟧ = ⟦⟨μ, τ⟩⟧) ; ⟨N steps with μ, τ⟩
⟦untime τ⟧$_t$ = ⟦τ⟧$_t$; 'removeClock [none] { empty }

(d) General timed strategies.

Fig. 1. Interpretation of real-time strategies as Maude's strategies.

strategy is used to solve rewrite expressions in conditional rules). Maude's strategy all nondeterministically chooses, and applies once, one of the *executable* rewrite rules. Therefore, when $\mu = $ action, only executable *instantaneous* rules are applied. The strategy [apply [$\mathcal{L}$]] tries, in order, the instantaneous rules in the list $\mathcal{L}$. When $\mu = $ delay, the strategy $[\tau]_t$ is executed. The normalization operator all ! applies all until it cannot be further applied. Hence, when $\mu = $ eager, all the instantaneous transitions are (nondeterministically) applied as much as possible, followed by a tick, if possible. The interpretation of the strategies _;_, _or_, _or-else_ and if_then_else uses the corresponding constructors in Maude's strategy language. In the case $\mu = $ get M' and set M'', Maude's matchrew is used to bind M' with the needed entries in the map storing information about the execution of the strategy. Then, the execution of the rule updateMap replaces the values in M' with the corresponding ones in M''.

5.4 User-Friendly Analysis Commands

A user-defined strategy $\langle \mu, \tau \rangle$ controls "one round" of the execution of the system. In this section we provide convenient "Real-Time Maude-like" syntax for most simulation, reachability and other formal analysis methods provided by Real-Time Maude, albeit executed with user-defined strategies. For that, user-defined strategies and user-defined time-sampling strategies are extended as follows.

Discrete strategies (of sort DStrat), besides the basic user-defined discrete strategies, include: the strategy check ϕ that fails if ϕ does not hold in the current state; the conditional repetition of a given strategy until ϕ do μ ; the strategy repeat μ that iteratively executes μ until it fails; and the strategy n steps with μ that repeats n times μ. Overloaded operators for the sort DStrat (e.g., op _;_ ... [ditto]) are also defined and omitted here.

```
sort DStrat .      subsort UDStrat < DStrat .
op check_            : SCond               -> DStrat .
op until_do_         : SCond DStrat        -> DStrat .
op repeat_           : DStrat              -> DStrat .
op _steps with_      : Nat DStrat          -> DStrat .
```

General timed strategies (of sort TStrat) extend user-defined time sampling strategies with a new case, used later to define untimed reachability analysis:

```
sort TStrat .      subsort UTStrat < TStrat .
op untime  : TStrat -> TStrat .
```

untime τ applies τ and then the rule removeClock, thus removing the global clock from the current state. These new strategy constructors are defined as Maude strategies as shown in Fig. 1d.

Commands. We define a convenient syntax for Real-Time Maude-like analysis commands using strategies. Given a user-defined strategy $\langle \mu, \tau \rangle$, we define a strategy $\langle \mu', \tau' \rangle$ that implements such an analysis command by rewriting (using metaSrewrite) an initial state *init* and returning a list of ClockedSystems (the solutions). We refer to [25] for our time-bounded simulation command.

Unbounded and time-bounded reachability commands are defined as:

```
op tsearch [_] in_:_=>_using_with sampling_ :
    Nat Qid StrState SCond DStrat TStrat -> LClockedSystem .
op tsearch [_] in_:_=>_using_with sampling_in time_ :  ... -> ... .
```

`tsearch [n] in R : init => `ϕ` using `μ` with sampling `τ
returns the first n states that result from *init* by rewriting with the strategy $\langle$`repeat `μ` ; check `$\phi, \tau\rangle$, i.e., repeat μ zero or more times and, on the resulting term, check ϕ. Time-bounded reachability analysis `tsearch [n] in R : init => `ϕ` using `μ` with`
`sampling `τ` in time [a,b]` is implemented as the extended strategy
$\langle$`repeat (if after(`b`) then stop else `μ`) ; check(`$\phi \wedge$` in [`a,b`]), `$\tau\rangle$.

"Depth-bounded" versions of the form `tsearch [n,d] ...` of the above commands are also available. Furthermore, similar commands `dsearch` are defined where `metaSrewrite` is invoked with the flag `depthFirst`, thus exploring the rewriting graph in a "depth-first" manner.

Untimed reachability analysis is possible with the command

```
op usearch [_] in_:_=>_using_with sampling_ :
    Nat Qid StrState SCond DStrat TStrat -> LClockedSystem .
```

The implementation of this command is similar to the one for `tsearch` but the sampling strategy used is `untime(`τ`)`: after each tick, the global clock is removed from the state. A depth-first version `dusearch` is also available.

Finding the longest and the shortest time it takes to reach a desired state is supported by the following commands:

```
op find latest in_:_=>_using_with sampling_ :
    Qid StrState SCond DStrat TStrat                   -> LClockedSystem .
op find earliest in_:_=>_using_with sampling_ :  ... -> LClockedSystem .
```

`find latest` uses `metaSrewrite` to find *all* the solutions when rewriting the initial state with the strategy $\langle$`until `ϕ` do `μ` ; check `ϕ` , `$\tau\rangle$. This finds the *first* state in *all* the branches of the search tree that satisfies ϕ. We then post-process the returned list to find the state with the greatest global clock value. This command may not terminate if there is a branch where ϕ never holds.

For `find earliest`, let t be the global clock value in the first state found when applying the strategy $\langle$`until `ϕ` do `μ` ; check `ϕ` , `$\tau\rangle$. Then, the strategy $\langle$`until `ϕ` do (if after(`t`) then stop else `μ`) ; check `ψ` , `$\tau\rangle$, where $\psi = \phi \wedge$ `before(`t`)`, is applied to find a new solution whose global clock is strictly smaller than t. This procedure is repeated until no further solutions are found.

5.5 Example: Analyzing the Round Trip Time Protocol

We illustrate the use of our timed strategy language on the RTT example; additional examples are given in our longer report [25].

We check whether it is possible to reach two states with RTT value 20 with the optimal "mixed" time sampling strategy that visits each time instant when there is a message in the system, and uses maximal time sampling otherwise:

```
Maude> red tsearch [2] in 'RTT : init =>
       matches ({CONF < S : Sender | rtt : 20, ATTS >} in time R:Time)
       using delay or action with sampling
           (switch when matches ({CONF dly(M, T1, T2)} in time R':Time)
           do fixed-time 1 otherwise max-time with default 1) .
result NeList{ClockedSystem}:
  ({< snd : Sender | rtt : 20, ... > ... } in time 20)
  ({< snd : Sender | rtt : 20, ... > ... } in time 5000)
```

Time-bounded search for RTT 20 with maximal time sampling finds no solution:

```
Maude> red tsearch [2] ... => matches ... rtt : 20 ... using delay or
       action with sampling max-time with default 4 in time [5000, 10000] .
result LClockedSystem: (nil).LClockedSystem  --- (No solution)
```

We then check the longest and shortest time needed to record an RTT value different from 0 and INF (rtt?(STATE)) for the first time in each behavior:

```
Maude> red find earliest in 'RTT : init => matches ({STATE} in time T2)
       s.t. (rtt?(STATE)) using action or delay with sampling fixed-time 1 .
result ClockedSystem: {< snd : Sender | rtt : 12, ... > ... } in time 12

Maude> red find latest in 'RTT : ... with sampling fixed-time 1 .
result ClockedSystem: {< snd : Sender | rtt : 50, > ... } in time 50
```

Let μ be the history-dependent strategy in Example 2 using the rule skipRound. This strategy covers all interesting behaviors (avoiding the second round of RTT) and it explores 126 states in the interval [0,10000]:

```
Maude> red size(tsearch in 'RTT : init => matches ({STATE} in time T2)
       using μ with sampling max-time with default 4 in time [0, 100000]) .
result NzNat: 126
```

A similar command using the strategy action or delay reports 162 states.

5.6 Benchmarking

We compare the performance of our strategy-based analysis methods with standard Maude search on a variation of the CASH scheduling algorithm developed by Marco Caccamo at UIUC [11]. The idea of CASH is that some jobs may not need all the execution times allocated to them. These unused clock cycles are put in a *queue* for other jobs to use. CASH is a sophisticated algorithm, with sporadic tasks (i.e., a job could arrive at any time), unknown length of each job, and a queue of unused execution times. Real-Time Maude analysis discovered the previously unknown fact that hard deadlines could be missed [28].

We transform the Real-Time Maude specification of the CASH protocol into a "standard" Maude model by incrementing time by one unit in the tick rules (see [24]). We can therefore use Maude's `search` command to find whether it is possible to reach a state where a deadline is missed within time 12:[4]

```
Maude> search [1] init =>* {DEADLINE-MISS CONF} in time T s.t. T <= 12 .

Solution 1 (state 599272)
rewrites: 34093729 in 14910ms cpu (14937ms real) ...
```

The `tsearch` and `dtsearch` (depth-first search) commands in our language that correspond to this time-bounded reachability query are executed as follows:

```
Maude> red tsearch [1] in 'CASH : init =>
       matches ({DEADLINE-MISS CONF} in time T)
       using delay or action with sampling fixed-time 1 in time [0, 12] .

rewrites: 44 in 19517ms cpu (19538ms real) ...

Maude> red dtsearch [1] ... in time [0, 12] .

rewrites: 44 in 2079ms cpu (2083ms real) ...
```

We also perform *unbounded* reachability analysis with Maude's `search`, and the commands `tsearch` and `usearch`. (`dtsearch` in this case did not terminate). Maude's `search` command, without constraints on the system clock, finds the missed deadline in **15 s**. Unbounded `tsearch` needed **22 s**, and `usearch`, which removes the system clock after each tick step, needed **9.8 s**. For an even more optimized Maude `search`, we also modified our Maude specification by *manually* removing the "in time ..." part of each tick rule, so that the state does not carry the system clock. In that case, Maude `search` terminates in **4.5 s**.

All experiments were run on a Dell XPS 13 laptop (with an Intel i7 processor @ 1.30GHz and 16GB of RAM). For time-bounded reachability, `tsearch` (19.5 s) is not much slower than Maude's `search` (15 s) on a Maude model where the deterministic time sampling strategy with increment 1 is hard-coded in the tick rule (and no meta-level procedure is used as in `tsearch`). Furthermore, our "depth-first" search command `dtsearch` significantly outperforms Maude's `search` command on this application (2 s).

In the unbounded case, it is fair to compare `tsearch` (22 s), which carries the system clock, to the Maude `search` which took 15 s, and `usearch` (9.8 s) to the Maude `search` of the manually modified model without the system clock (4.5 s). All in all, our strategy-implemented commands are reasonably close to Maude `search`, and in one case even significantly faster.

Because of the massive time-nondeterminism in the model (jobs can arrive at any time and may execute for an any amount of time), we must use deterministic time sampling with increment 1 for CASH. The performance should be much better on systems such as RTT where we should use mixed time sampling to "ignore" idling states where not much can happen. Although the preliminary results are promising, we should do more thorough benchmarking in future work.

---

[4] Here `init` denotes an initial state from which a missed deadline should not be reachable if the optimized version of CASH were correct.

6 Related Work

UPPAAL STRATEGO [14] extends the timed automaton tool UPPAAL [7] with strategies and model checking under such strategies, where strategies are UPPAAL queries. UPPAAL STRATEGO seems to be used mainly in connection with synthesis of controller strategies. We target a more expressive formalism, provide a language for specifying actual strategies instead of queries, and also provide time sampling strategies, but we do not support synthesizing strategies.

Different strategy languages have been proposed to cope with the nondeterminism in rewriting. Examples of such languages include ELAN [9], Stratego [10], and ρLog [21]. Applications of Maude's strategy language [15] include the analysis neural networks [39], membrane systems [37], and the specification of semantics of programming languages [17] and process calculi [34]. Our previous work [2–4] represents, to the best of our knowledge, the first applications of the Maude strategy language to (simple) real-time systems. Those efforts motivated the development of the intuitive timed strategy language proposed here.

The paper [6] uses rewrite rules and "strategies" to analyze timed automata reachability using the rewriting framework ELAN [9]. The authors define rewrite rules for manipulating "zones" of the timed automaton, and then define rewrite strategies for various approaches to analyze these symbolic state spaces, whereas we use strategies to explore subsets of system behaviors.

7 Concluding Remarks

In this paper we propose what we hope is a useful yet reasonably intuitive language for defining execution strategies for real-time systems in Maude, allowing us to perform most of the analysis methods supported by Real-Time Maude, with user-defined discrete and timed strategies. We identify a number of interesting execution strategies for real-time systems, including a "mixed" time sampling strategy that should be ideal for explicit-state analysis of a large class of distributed real-time systems, such as our round trip time protocol.

Our strategies are given a semantics in Maude and are therefore implemented in Maude. A preliminary performance comparison between standard Maude search and our strategy-implemented reachability analyses on the CASH scheduling algorithm benchmark indicates that the latter are fairly competitive.

The benefits of this work are: (i) allowing the user to quickly and easily analyze her real-time system under a wide range of different scenarios without having to modify her model; (ii) providing much better time sampling strategies for time-sampling-based explicit-state analysis than those provided by Real-Time Maude; (iii) providing a convenient framework for quickly experimenting with different strategies and analyses, before optimizing and hard-coding the most promising into the Real-Time Maude tool; (iv) allowing us to analyze real-time rewrite theories directly in Maude, instead of in Real-Time Maude; and (v) supporting formal analysis with user-defined strategies for modeling languages and formalisms for which Real-Time Maude provides a formal analysis backend.

Rubio et al. [35,36] have shown how to model check strategy-aware rewriting logic specifications in their umaudemc tool, which allows model checking LTL and CTL formulas, as well as to perform probabilistic and statistical model-checking, on systems controlled by strategies. In future work we should support untimed and timed temporal logic model checking combined with real-time strategies. We should also combine *symbolic* analysis of real-time rewrite theories with user-defined strategies, as we did in [2,3] for timed automata and Petri nets.

Acknowledgments. This work was supported by the NATO Science for Peace and Security Programme through grant number G6133 (project SymSafe) and by the PHC project Aurora AESIR.

References

1. AlTurki, M., Dhurjati, D., Yu, D., Chander, A., Inamura, H.: Formal specification and analysis of timing properties in software systems. In: Fundamental Approaches to Software Engineering (FASE 2009). LNCS, vol. 5503, pp. 262–277. Springer, Berlin (2009). https://doi.org/10.1007/978-3-642-00593-0_18
2. Arias, J., Bae, K., Olarte, C., Ölveczky, P.C., Petrucci, L., Rømming, F.: Rewriting logic semantics and symbolic analysis for parametric timed automata. In: Proceedings of the 8th ACM SIGPLAN International Workshop on Formal Techniques for Safety-Critical Systems (FTSCS 2022), pp. 3–15. ACM (2022). https://doi.org/10.1145/3563822.3569923
3. Arias, J., Bae, K., Olarte, C., Ölveczky, P.C., Petrucci, L., Rømming, F.: Symbolic analysis and parameter synthesis for time Petri nets using Maude and SMT solving. In: Application and Theory of Petri Nets and Concurrency (PETRI NETS 2023). LNCS, vol. 13929, pp. 369–392. Springer, Berlin (2023). https://doi.org/10.1007/978-3-031-33620-1_20
4. Arias, J., Bae, K., Olarte, C., Ölveczky, P.C., Petrucci, L., Rømming, F.: Symbolic analysis and parameter synthesis for networks of parametric timed automata with global variables using Maude and SMT solving. Sci. Comput. Program. **233** (2024).https://doi.org/10.1016/j.scico.2023.103074
5. Bae, K., Ölveczky, P.C., Feng, T.H., Lee, E.A., Tripakis, S.: Verifying hierarchical Ptolemy II discrete-event models using Real-Time Maude. Sci. Comput. Program. **77**(12), 1235–1271 (2012). https://doi.org/10.1016/j.scico.2010.10.002
6. Beffara, E., Bournez, O., Kacem, H., Kirchner, C.: Verification of timed automata using rewrite rules and strategies (2009). https://doi.org/10.48550/arXiv.0907.3123
7. Behrmann, G., David, A., Larsen, K.G.: A tutorial on Uppaal. In: Formal Methods for the Design of Real-Time Systems (SFM-RT 2004). LNCS, vol. 3185, pp. 200–236. Springer, Berlin (2004). https://doi.org/10.1007/978-3-540-30080-9_7
8. Bobba, R., Grov, J., Gupta, I., Liu, S., Meseguer, J., Ölveczky, P.C., Skeirik, S.: Survivability: Design, formal modeling, and validation of cloud storage systems using Maude. In: Assured Cloud Computing, Chap. 2, pp. 10–48. Wiley, New York (2018). https://doi.org/10.1002/9781119428497.ch2
9. Borovanský, P., Kirchner, C., Kirchner, H., Ringeissen, C.: Rewriting with strategies in ELAN: A functional semantics. Int. J. Found. Comput. Sci. **12**(1), 69–95 (2001). https://doi.org/10.1142/S0129054101000412

10. Bravenboer, M., Kalleberg, K.T., Vermaas, R., Visser, E.: Stratego/XT 0.17. A language and toolset for program transformation. Sci. Comput. Program. **72**(1–2), 52–70 (2008). https://doi.org/10.1016/J.SCICO.2007.11.003
11. Caccamo, M., Buttazzo, G.C., Sha, L.: Capacity sharing for overrun control. In: Proceedings of the 21st IEEE Real-Time Systems Symposium (RTSS 2000), pp. 295–304. IEEE Computer Society (2000). https://doi.org/10.1109/REAL.2000.896018
12. Clavel, M., Durán, F., Eker, S., Escobar, S., Lincoln, P., Martí-Oliet, N., Meseguer, J., Rubio, R., Talcott, C.: Maude Manual (Version 3.3.1). SRI International (2023). available at http://maude.cs.illinois.edu
13. Clavel, M., Durán, F., Eker, S., Lincoln, P., Martí-Oliet, N., Meseguer, J., Talcott, C.L.: All About Maude—A High-Performance Logical Framework, LNCS, vol. 4350. Springer, Berlin (2007) https://doi.org/10.1007/978-3-540-71999-1
14. David, A., Jensen, P.G., Larsen, K.G., Mikucionis, M., Taankvist, J.H.: Uppaal Stratego. In: TACAS 2015. LNCS, vol. 9035. Springer, Berlin (2015). https://doi.org/10.1007/978-3-662-46681-0_16
15. Eker, S., Martí-Oliet, N., Meseguer, J., Rubio, R., Verdejo, A.: The Maude strategy language. J. Log. Algebraic Methods Program. **134**, 100887 (2023). https://doi.org/10.1016/J.JLAMP.2023.100887
16. Grov, J., Ölveczky, P.C.: Formal modeling and analysis of Google's Megastore in real-time Maude. In: Specification, Algebra, and Software—Essays Dedicated to Kokichi Futatsugi. LNCS, vol. 8373, pp. 494–519. Springer, Berlin (2014). https://doi.org/10.1007/978-3-642-54624-2_25
17. Hidalgo-Herrero, M., Verdejo, A., Ortega-Mallén, Y.: Using Maude and its strategies for defining a framework for analyzing Eden semantics. In: Antoy, S. (ed.) WRS@FLoC 2006. ENTCS, vol. 174, pp. 119–137. Elsevier (2006). https://doi.org/10.1016/J.ENTCS.2007.02.051
18. Lepri, D., Ábrahám, E., Ölveczky, P.C.: Sound and complete timed CTL model checking of timed Kripke structures and real-time rewrite theories. Sci. Comput. Program. **99**, 128–192 (2015). https://doi.org/10.1016/j.scico.2014.06.006
19. Liu, S., Meseguer, J., Ölveczky, P.C., Zhang, M., Basin, D.A.: Bridging the semantic gap between qualitative and quantitative models of distributed systems. Proc. ACM Program. Lang. **6**(OOPSLA2), 315–344 (2022). https://doi.org/10.1145/3563299
20. Liu, S., Ölveczky, P.C., Meseguer, J.: Modeling and analyzing mobile ad hoc networks in Real-Time Maude. J. Log. Algebraic Methods Program. **85**(1), 34–66 (2016). https://doi.org/10.1016/j.jlamp.2015.05.002
21. Marin, M., Kutsia, T.: Foundations of the rule-based system rLog. J. Appl. Non Class. Logics **16**(1–2), 151–168 (2006). https://doi.org/10.3166/JANCL.16.151-168
22. Meseguer, J.: Conditional rewriting logic as a unified model of concurrency. Theor. Comput. Sci. **96**(1), 73–155 (1992). https://doi.org/10.1016/0304-3975(92)90182-F
23. Meseguer, J.: Membership algebra as a logical framework for equational specification. In: Recent Trends in Algebraic Development Techniques (WADT'97). LNCS, vol. 1376, pp. 18–61. Springer, Berlin (1997).https://doi.org/10.1007/3-540-64299-4_26
24. Olarte, C., Ölveczky, P.C.: RT-Strategies (2024). https://depot.lipn.univ-paris13.fr/real-time-maude/rt-strategies.git
25. Olarte, C., Ölveczky, P.C.: Timed strategies for real-time rewrite theories (2024). https://arxiv.org/abs/2403.08920

26. Ölveczky, P.C.: Real-Time Maude and its applications. In: Rewriting Logic and Its Applications (WRLA 2014). LNCS, vol. 8663, pp. 42–79. Springer, Berlin (2014). https://doi.org/10.1007/978-3-319-12904-4_3
27. Ölveczky, P.C., Boronat, A., Meseguer, J.: Formal semantics and analysis of behavioral AADL models in Real-Time Maude. In: Formal Techniques for Distributed Systems, Joint 12th IFIP WG 6.1 International Conference, FMOODS 2010 and 30th IFIP WG 6.1 FORTE 2010. LNCS, vol. 6117, pp. 47–62. Springer, Berlin (2010). https://doi.org/10.1007/978-3-642-13464-7_5
28. Ölveczky, P.C., Caccamo, M.: Formal simulation and analysis of the CASH scheduling algorithm in Real-Time Maude. In: Fundamental Approaches to Software Engineering (FASE 2006). LNCS, vol. 3922, pp. 357–372. Springer, Berlin (2006). https://doi.org/10.1007/11693017_26
29. Ölveczky, P.C., Meseguer, J.: Specification of real-time and hybrid systems in rewriting logic. Theor. Comput. Sci. **285**(2), 359–405 (2002). https://doi.org/10.1016/S0304-3975(01)00363-2
30. Ölveczky, P.C., Meseguer, J.: Semantics and pragmatics of Real-Time Maude. High. Order Symb. Comput. **20**(1–2), 161–196 (2007). https://doi.org/10.1007/s10990-007-9001-5
31. Ölveczky, P.C., Meseguer, J.: The Real-Time Maude tool. In: Tools and Algorithms for the Construction and Analysis of Systems (TACAS 2008). LNCS, vol. 4963, pp. 332–336. Springer, Berlin (2008). https://doi.org/10.1007/978-3-540-78800-3_23
32. Ölveczky, P.C., Meseguer, J., Talcott, C.L.: Specification and analysis of the AER/NCA active network protocol suite in Real-Time Maude. Formal Methods Syst. Des. **29**(3), 253–293 (2006). https://doi.org/10.1007/s10703-006-0015-0
33. Ölveczky, P.C., Thorvaldsen, S.: Formal modeling, performance estimation, and model checking of wireless sensor network algorithms in Real-Time Maude. Theor. Comput. Sci. **410**(2–3), 254–280 (2009). https://doi.org/10.1016/j.tcs.2008.09.022
34. Rosa-Velardo, F., Segura, C., Verdejo, A.: Typed mobile ambients in Maude. In: Cirstea, H., Martí-Oliet, N. (eds.) RULE@RDP 2005. ENTCS, vol. 147, pp. 135–161. Elsevier (2005). https://doi.org/10.1016/J.ENTCS.2005.06.041
35. Rubio, R., Martí-Oliet, N., Pita, I., Verdejo, A.: Strategies, model checking and branching-time properties in Maude. J. Log. Algebraic Methods Program. **123**, 100700 (2021). https://doi.org/10.1016/J.JLAMP.2021.100700
36. Rubio, R., Martí-Oliet, N., Pita, I., Verdejo, A.: Model checking strategy-controlled systems in rewriting logic. Autom. Softw. Eng. **29**(1), 7 (2022). https://doi.org/10.1007/S10515-021-00307-9
37. Rubio, R., Martí-Oliet, N., Pita, I., Verdejo, A.: Simulating and model checking membrane systems using strategies in Maude. J. Log. Algebraic Methods Program. **124**, 100727 (2022). https://doi.org/10.1016/J.JLAMP.2021.100727
38. Sabahi-Kaviani, Z., Khosravi, R., Ölveczky, P.C., Khamespanah, E., Sirjani, M.: Formal semantics and efficient analysis of Timed Rebeca in Real-Time Maude. Sci. Comput. Program. **113**, 85–118 (2015). https://doi.org/10.1016/J.SCICO.2015.07.003
39. Santos-García, G., Palomino, M., Verdejo, A.: Rewriting logic using strategies for neural networks: An implementation in Maude. In: Corchado, J.M., Rodríguez, S., Llinas, J., Molina, J.M. (eds.) DCAI 2008. Advances in Soft Computing, vol. 50, pp. 424–433. Springer, Berlin (2008). https://doi.org/10.1007/978-3-540-85863-8_50

Specifying Fairness Constraints and Model Checking with Non-intensional Strategies

Rubén Rubio[✉], Narciso Martí-Oliet, Isabel Pita, and Alberto Verdejo

Facultad de Informtica, Universidad Complutense de Madrid, Madrid, Spain
{rubenrub,narciso,ipandreu,jalberto}@ucm.es

Abstract. Strategies are a natural way of specifying constraints in a rewriting-based model. They are often expressed as executable programs in some strategy language that intensionally filter the possible next rewrites. Hence, they cannot capture restrictions of the model involving unbounded delays, like fairness, which are useful in many verification scenarios. In this paper, we propose a variation to the semantics of the Maude strategy language that allows expressing non-intensional strategies without recursion, in a way amenable for verification. Then we present an LTL model checker for this kind of strategies and discuss the corresponding problem for other logics.

1 Introduction

In order to formally reason about discrete systems, their behavior and properties should be conveniently modeled. State and transition systems are the simplest model of a dynamic system, yet the most useful for several verification techniques. For model checking [10], transition systems are extended by labeling their states or transitions with atomic propositions, which are then used to specify properties about their dynamic behavior using temporal logics like LTL, CTL, and μ-calculus. While transition systems are adequate to represent the local, executable behavior of the system, they cannot embody whole-execution restrictions involving unbounded nondeterminism, like fairness, which are convenient in several verification scenarios. In logics like LTL, fairness constraints can be expressed in the formula itself as long as the atomic propositions are detailed enough. For some other logics like CTL, model-checking algorithms have been extended to receive fairness constraints as another piece of input. Moreover, some formalisms have been proposed to embed those restrictions into the model itself, like fair Kripke structures [20].

Maude [11] is a high-level specification and programming language based on rewriting logic [23]. Specifications are organized in modules: functional modules describe membership equational theories [5] with sort and operator declarations, equations, and membership axioms; system modules extend them with rewrite rules acting on those terms; and finally, strategy modules define strategies to control or guide the application of rules using the Maude strategy language [13,16]. Maude specifications, even strategy-controlled ones [30,31], can be seen as Kripke structures with the terms as states and their rewrites as transitions, and model checked with a built-in LTL model checker [17] or other external tools [1,2,30,31]. In this context, strategies are useful to represent constraints on the model behavior, but they cannot capture requirements entailing unbounded nondeterminism with the usual executable semantics [12]. The strategy-constrained model used for model checking is also a standard Kripke structure that cannot represent fairness or similar properties.

Several combinators of the Maude strategy language (and other similar languages like ELAN [4], Stratego [7], ρLog [21], or Porgy [18]) are inspired on regular expressions: union (|), concatenation (;), the empty language (`fail`), the empty word (`idle`), basic symbols (rule applications), and the Kleene star (*). Like in regular expressions, the Kleene star of the Maude strategy language (called *iteration*) executes its argument zero or more times nondeterministically, but unlike them, it also allows repeating its argument forever. In other words, $\alpha*$ in the strategy language corresponds to $\alpha^* \mid \alpha^\omega$ instead of α^* as an ω-regular expression. This deviation is completely immaterial when evaluating strategies to compute results, but it may matter when model checking. Using a faithful semantics for the Kleene star operator, we will be able to represent fairness and similar constraints within the strategy, at the expense of some additional complexity.

In this paper, we modify the semantics of the Maude strategy language in order to maintain the usual meaning of the Kleene star and so allow expressing long-run constraints within strategies. Model checking under this interpretation of strategies is achieved by extending the standard automata-theoretic approach for LTL model checking [10]. In the standard case, in order to check $\mathcal{M} \vDash \varphi$, the transition system $\mathcal{M}$ is seen as a trivial automaton, the negated LTL property $\neg\varphi$ is transformed to a Büchi automaton, and the emptiness of their product is checked. If the product is empty, the property is satisfied; and otherwise, an accepted run in the product automaton yields an execution that does not satisfy the temporal property. In our case, instead of a plain transition system $\mathcal{M}$, we generate a Streett automaton for the strategy-controlled Maude model and proceed likewise. This procedure is implemented in our tool umaudemc [29], which relies on the Maude Python bindings [26] to communicate with Maude and on the Spot library [15] to build the ω-automata and operate with them. For other logics, model checking can also be achieved by generating artificial atomic propositions and transforming the input formula, an approach that is also implemented in our tool.

The paper is organized as follows. After some preliminaries in Sect. 2, the standard small-step semantics of the Maude strategy language is explained in Sect. 3 and modified as anticipated before in Sect. 4. Section 5 discusses how to model check LTL properties for strategy-controlled systems under this semantics, explains the implementation details, and outlines an alternative method amenable for other logics. Finally, Sect. 6 discusses related work and Sect. 7 concludes the paper. This paper is based on unpublished material from the first author's PhD thesis [28], although significantly improved with a new implementation and simpler formalizations. Source code, documentation, and examples can be found at https://www.github.com/fadoss/umaudemc.

2 Preliminaries

In this section, we recall some topics and introduce some notation that will be used in the rest of the paper. Given an alphabet S, a finite word $s_1 \cdots s_n \in S^*$ is a finite sequence of symbols in S, and ε denotes the empty word. Similarly, an infinite word $s_0 s_1 \cdots \in S^\omega$ is an infinite sequence of symbols in S. We write w_i for the i-th symbol in w, and $w^i = w_i w_{i+1} \cdots$. For an infinite word $w \in S^\omega$, we denote by $\inf(w)$ the set of all symbols $s \in S$ that appear infinitely often in w.

Strategies. Since rewrite rules can be applied in different orders and into different subterms, rewriting is intrinsically nondeterministic. However, sometimes it is convenient to control the application of rules to exclude unwanted behaviors or guide a search to the desired goal. Rewriting strategies play that role and they have been widely studied in the context of functional programming, the λ-calculus, and abstract rewriting. In [6], the authors propose two convenient abstract formalization of strategies. Given an abstract reduction system $\mathcal{A} = (S, \to)$ with a set of states S and a binary relation $\to$, its set of executions is $\mathrm{Ex}^\omega(\mathcal{A}) = \{\pi \in S^\omega \mid \pi_k \to \pi_{k+1}, k \in \mathbb{N}\}$.[1] Strategies can be described

- *extensionally* as a subset of executions $E \subseteq \mathrm{Ex}^\omega(\mathcal{A})$ of the system, or
- *intensionally* as partial functions $\lambda : S^* \to \mathcal{P}(S)$ specifying the possible next states $\lambda(w)$ to continue each partial execution w.

Extensional strategies are strictly more general than intensional ones, because for any λ there is $E(\lambda) := \{\pi \in S^\omega \mid \pi_k \in \lambda(\pi_0 \cdots \pi_{k-1}), k \in \mathbb{N}\}$, and $\{a^n b^\omega \mid n \in \mathbb{N}\}$ cannot be represented intensionally (a^ω cannot be excluded). Indeed, the extensional denotations of intensional strategies are characterized as being *closed*, i.e. $\pi \in S^\omega$ is allowed by the strategy if infinitely many prefixes of π are prefixes of words allowed by the strategy [6, Proposition 3]. In the following, we will denote strategies as subsets of executions, i.e. in the extensional way.

Model checking. Model checking [10] is an automated verification technique where models are expressed as transition systems and properties as formulas in some temporal logic. A Kripke structure is a tuple $(S, \to, I, AP, \ell)$ where S is a set of states, $(\to) \subseteq S \times S$ is a transition relation, $I \subseteq S$ is a subset of initial states, AP is a finite set of atomic propositions, and $\ell : S \to \mathcal{P}(AP)$ is the labeling function of states by atomic propositions. These atomic propositions $a \in AP$ are combined with logical and temporal operators to build formulae that describe the intended behavior of the system. Linear-time Temporal Logic

---

[1] For simplicity, only nonterminating executions are considered, as usual in the model-checking literature. Finite ones can be extended by repeating the last state forever.

(LTL) [25] includes operators like $\bigcirc \varphi$ to tell that φ holds in the next state, $\square \varphi$ meaning that φ always holds, $\Diamond \varphi$ when φ holds at some point in the future, and $\varphi_1 \mathbf{U} \varphi_2$ to tell that φ_2 holds at some point in the future and φ_1 holds until then. Other widespread temporal logics are CTL, CTL*, and the μ-calculus.

Maude includes a built-in LTL model checker [16], which has been applied to several interesting problems [22]. As already mentioned, Maude specifications can be seen as Kripke structures by taking the terms as states and their rewrites as the transitions of the model. Atomic propositions can also be represented as terms and their satisfaction be established equationally. In previous works [30, 31], we have extended the model checker to work with specifications controlled by strategies in the Maude strategy language. Essentially, regarding a strategy as a subset of executions, a temporal property φ holds under a strategy E if it holds for all executions in E (or in the tree of executions pruned by E for branching-time logics like CTL).

Automata over infinite words. Many LTL model checkers use the so-called *automata-theoretic approach*, which reduces model checking to a language containment problem $\ell(\mathrm{Ex}^\omega(\mathcal{K})) \subseteq L(\varphi)$ that can be solved with automata. A Büchi automaton [10] is a tuple $(Q, \Sigma, \delta, Q_0, F)$ where Q is a finite set of states, Σ is a finite alphabet, $\delta : Q \times \Sigma \to \mathcal{P}(Q)$ is a nondeterministic transition function, $Q_0 \subseteq Q$ is a set of initial states, and $F \subseteq Q$ is an acceptance condition. $\pi \in Q^\omega$ is a *run* for a word $w \in \Sigma^\omega$ if $\pi_0 \in Q_0$ and $\pi_{k+1} \in \delta(\pi_k, w_k)$. The run π is *accepting* (then w is *accepted* by the automaton) if $\inf(\pi) \cap F \neq \emptyset$, i.e. if some state in F is repeated infinitely often. Well-known procedures exist to translate LTL formulae to Büchi automata [19].

Sometimes, deciding acceptance based on the transitions instead of the states is more convenient, and this is what the Spot library does by default. Moreover, the way of expressing acceptance conditions can be generalized. A transition-based Streett automaton [14] of index n is a tuple $(Q, \Sigma, Acc, \delta, Q_0, F)$ where Acc is a finite set of acceptance conditions, $\delta : Q \times \Sigma \to \mathcal{P}(Q \times \mathcal{P}(Acc))$, and $F = \{(a_1, b_1), \ldots, (a_n, b_n)\} \subseteq Acc^2$. Similarly, a run is defined as a sequence of pairs $Q \times \mathcal{P}(Acc)$, and $(q_i, A_i)_{i=0}^\infty$ is accepting if $\{i \in \mathbb{N} \mid a_k \in A_i\}$ is finite or $\{i \in \mathbb{N} \mid b_k \in A_i\}$ is not finite for all $1 \leq k \leq n$.

Coming back to model checking, a Kripke structure $(S, \to, I, AP, \ell)$ can be transformed to an equivalent Büchi automaton $(S, \mathcal{P}(AP), \delta, I, \emptyset)$ with $\delta(s, P) = \{s' \in S \mid s \to s'\}$ if $\ell(s) = P$, and $\delta(s, P) = \emptyset$ otherwise, both with trivial acceptance condition $F = \emptyset$. Then, the automata-theoretic approach proceeds by checking the emptiness of the product automaton between the system and the negated temporal formula. We will slightly adapt this approach to our purposes in the next sections.

3 The Maude Strategy Language and Its Semantics

The Maude strategy language [13,16] is intended to control the application of rules when executing and model checking Maude specifications. Its main instruction is then the application of a rule, written rl where rl is a rule label.[2] Several operators, similar to those of other strategy languages like ELAN [4], Tom [3], Stratego [7], ρLog [21], or Porgy [18], can be used to combine these rule applications into more complex strategic programs. In general, by applying a strategy expression α to a term t, we expand those rewriting paths from t that are *allowed* by the strategy. Expressions are built with the concatenation operator α ; β that continues rewriting with β every result of α, the nondeterministic choice α | β that allows any rewriting path allowed by α or β, fail discards the current execution path, idle simply continues to the next instruction, tests match P s.t. C are equivalent to idle when the term t matches P and satisfies C and to fail otherwise, and the conditional α ? β : γ that continues executing with β on the results of α, but executes γ directly from the initial term if α does not produce any result. Moreover, strategies can be applied into specific subterms using matchrew $P(x_1, \ldots, x_n)$ s.t. C by x_1 using α_1, ..., x_n using α_n, and recursive strategy definitions with arguments can be defined in strategy modules. Another useful combinator, the "star" of this paper, is the *iteration* strategy $\alpha*$ that executes α zero or more consecutive times nondeterministically. Under the standard semantics, it is recursively equivalent to idle | (α ; $\alpha*$).

Let us illustrate strategies with a simple example. The cups and balls is an ancient performance made with a table, three cups, and a small ball in one of its multiple variations. The illusionist puts the three cups upside down on the table, covering the marble with one of them, then swaps them randomly multiple times, and asks the audience to guess in which cup the ball is inside. The selected cup is raised to show whether the guess was right. The CUPS-BALLS module is a Maude system module defining some sorts (MaybeBall, Cup, Table), some constants and operators (ball, nothing, cup, ...), and finally some rules with labels swap, uncover, and cover. Moreover, the single argument of cup is declared frozen so that rules like cover cannot be applied on it, because inserting a cup below a cup does not make sense.

```
mod CUPS-BALLS is
    sorts MaybeBall Cup Table .
    subsorts MaybeBall Cup < Table .

    ops ball nothing :  -> MaybeBall [ctor] .
    op  cup     : MaybeBall -> Cup [ctor frozen] .
    op  empty   : -> Table [ctor] .
    op  __      : Table Table -> Table [ctor assoc id: empty] .

    var  T           : Table .
    vars B? B1? B2?  : MaybeBall .

    rl [swap]    : cup(B1?) T cup(B2?) => cup(B2?) T cup(B1?) .
    rl [uncover] : cup(B?) => B? .
    rl [cover]   : B? => cup(B?) .
```

---

[2] In this presentation, we will omit for clarity some details of the strategy language that are orthogonal to the contributions of the paper. For example, there are additional options in the rule application syntax that we are omitting here.

```
      op initial : -> Table .
      eq initial = cup(nothing) cup(ball) cup(nothing) .
endm
```

The illusionist will start with only a ball in the table, then it will cover the ball and the other two places with a cup, yielding our `initial` state. They will then swap the cups with `swap` for a while, one of them will be uncovered with `uncover` and shown to the public, then it will be covered back with `cover`, and the spectacle starts all over again. We want actions ordered this way, so we need strategies to limit the unwanted behaviors. For example, we do not want the illusionist to start swapping two cups before putting the third one down, as show in the following fragment of the uncontrolled rewrite graph.

```
Maude> search [, 2] initial =>* cup(ball) cup(nothing) nothing .
...
Maude> show search graph .
state 0, Table: cup(nothing) cup(ball) cup(nothing)
arc 5 ===> state 5 (uncover)
...
state 5, Table: cup(nothing) cup(ball) nothing
arc 1 ===> state 8 (swap)
...
state 8, Table: cup(ball) cup(nothing) nothing
```

The following strategy module `CUPS-BALLS-STRAT` provides a strategy `cups` that describes the previous procedure.

```
smod CUPS-BALLS-STRAT is    *** strategy module
    protecting CUPS-BALLS .

    strat cups @ Table .   *** strategy declaration
    sd cups := swap * ; uncover ; cover ; cups .   *** definition
endsm
```

Notice that `cups` is nonterminating. No term would be obtained as a result by rewriting with this strategy, but it can be used to describe the behavior of a long-running system for model checking.

In order to use the strategy-aware extension of the Maude model checker, we need to identify the sort of states and define some atomic propositions. This is done in the following module `CUPS-BALLS-PREDS`, which includes the `SATISFACTION` module provided by the Maude model checker [11, Sect. 12].

```
mod CUPS-BALLS-PREDS is
    protecting CUPS-BALLS .
    including SATISFACTION .

    subsort Table < State .
    ops uncovered hit : -> Prop [ctor] .

    vars L R T : Table .    var B? : MaybeBall .
```

```
       eq L B? R     |= uncovered = true .
       eq T          |= uncovered = false [owise] .
       eq L ball R   |= hit = true .
       eq T          |= hit = false [owise] .
endm
```

States are of sort Table and two atomic propositions are defined equationally on them: uncovered holds whenever the content of a cup (ball or nothing) is uncovered, and hit is satisfied when this content is ball. According to the usual meaning of the game, cups must be uncovered infinitely often, and this can be expressed in LTL as □ ◊ uncovered. However, if we check this with the strategy-aware model checker, we obtain the following:

```
$ umaudemc check cupsballs.maude initial '[] <> uncovered' cups
The property is not satisfied in the initial state
(13 system states, 52 rewrites, 2 Büchi states)
| cup(nothing) cup(ball) cup(nothing)
v    swap
| | cup(nothing) cup(nothing) cup(ball)
| v    swap
< v
```

The property is not satisfied because swap is repeated forever in a loop. While this is consistent with the meaning of the iteration, it is not probably what we want to express when we use the Kleene star.

3.1 Small-Step Operational Semantics

For the purpose of model checking, we introduced in previous works [30] a small-step operational semantics for the Maude strategy language, whose rules are shown in Fig. 1. The execution states $q \in \mathfrak{XS}$ of that semantics combine the term being rewritten with a strategy continuation,

$$q ::= t @ s \mid \text{subterm}(t; x : q, \ldots, x : q) @ s \qquad s ::= \varepsilon \mid \alpha\, s \mid \sigma\, s$$

where t is a term, x is a variable, α is a strategy expression (let Strat be the set of all strategy expressions), σ is a substitution (acting as a variable context), and s is a stack of strategies and substitutions. The subterm execution state holds the parallel rewriting states of the subterm rewriting operators (matchrew). For any state $q \in \mathfrak{XS}$, its current term cterm(q) is defined as cterm($t @ s$) = t and cterm(subterm($t; x_1 : q_1, \ldots, x_n : q_n$) @ s) = $t[x_1/\text{cterm}(q_1), \ldots, x_n/\text{cterm}(q_n)]$. The small-step operational semantics is defined by a twofold relation in Fig. 1: control transitions $\to_c$ advance the execution of the strategy without affecting the current term, while system transitions $\to_s$ perform a single rewrite in the current term. We write $\to_{s,c} = \to_s \cup \to_c$ and $\twoheadrightarrow = \to_s \circ \to_c^*$, i.e. a single rewrite preceded by as many control transitions as needed. It follows that $q \twoheadrightarrow q'$ implies that there is a rewrite from cterm(q) to cterm(q'). With these ingredients, we can define the set of rewriting paths allowed by an expression.

Definition 1. Given a strategy expression α and a term t,

$$E(\alpha, t) := \mathrm{cterm}(\mathrm{Ex}^\omega(\alpha, t))$$

where $\mathrm{Ex}^\omega(\alpha, t) := \{q_0(q_k)_{k=1}^\infty \mid q_0 = t\,@\,\alpha, q_k \twoheadrightarrow q_{k+1}\}$ are all nonterminating executions of α from t.

The semantic rule for strategy calls in Fig. 1 deserves some comment: for any matching strategy definition **sd** $sl(p_1, \ldots, p_n) := \delta$ in the current strategy module, it pushes the matching substitution σ of the call term into the left-hand side of the definition as variable context, and also the strategy expression δ to continue rewriting with it. Tail calls can be easily characterized as those in which the top of s is also a substitution, and in this case we replace the top substitution with the new one instead of preserving both. This allows expressing nonterminating cyclic executions with a finite state space, which is quite useful for model checking. This optimization can be seen as an equation $\sigma'\sigma = \sigma'$ on the execution stack.

We can then consider the Kripke structure $\mathfrak{M}_{\alpha,t} := (\mathfrak{XS}, \twoheadrightarrow, \{t\,@\,\alpha\}, AP, \ell')$ with $\ell'(q) := \ell(\mathrm{cterm}(q))$ if ℓ is the labelling function derived from the Maude specification. Since its executions are exactly $E(\alpha, t)$, we can apply standard model-checking algorithms on this structure to solve the strategy-controlled problem, as this is what the strategy-aware model checker does [30]. However, this semantics does not respect the usual meaning of the Kleene star, as we have seen in the previous section. Indeed, the procedure through $\mathfrak{M}_{\alpha,t}$ will not be possible with the Kleene star semantics.

4 A New Semantics that Respects the Kleene Star

The Kleene star usually designates all the finite concatenations of words taken from a given language, and this is the meaning it has in regular and ω-regular expressions. However, as we have just seen, the Maude strategy $\alpha*$ can be informally described as $\alpha^* \mid \alpha^\omega$, because it also admits repeating α forever. In effect, the following steps of the example in the previous section (where $\beta \equiv$ uncover ; cover ; cups)

$$\begin{aligned}
&\text{cup(nothing) cup(ball) cup(nothing) @ swap } *\ ;\beta \\
&\to_c \text{cup(nothing) cup(ball) cup(nothing) @ swap ; swap } *\ ;\beta \\
&\to_s \text{cup(nothing) cup(nothing) cup(ball) @ swap } *\ ;\beta \\
&\to_c \text{cup(nothing) cup(nothing) cup(ball) @ swap ; swap } *\ ;\beta \\
&\to_s \text{cup(nothing) cup(ball) cup(nothing) @ swap } *\ ;\beta
\end{aligned}$$

can be repeated indefinitely in a loop according to the semantics. For faithfully representing the Kleene star, we only need to remove those executions of a strategy where an iteration is repeated infinitely many times in a row. We formalize this with the following definition.

```
┌─ Rule applications and tests ─ ─ ─ ─ ─ ─ ─ ─ ─ ─ ─ ─ ─ ─ ─ ─ ─ ─ ─ ─ ─ ─ ─ ─ ─┐
│                                                                             │
│              $t \, @ \, rl \, s \rightarrow_s t' \, @ \, s$ if $t$ rewrites to $t'$ by $rl$                │
│                                                                             │
│       $t \, @ \, (\text{match } P \text{ s.t. } C) \, s \rightarrow_c t \, @ \, s$ if $t$ matches $P$ and satisfies $C$       │
│                                                                             │
└─ ─ ─ ─ ─ ─ ─ ─ ─ ─ ─ ─ ─ ─ ─ ─ ─ ─ ─ ─ ─ ─ ─ ─ ─ ─ ─ ─ ─ ─ ─ ─ ─ ─ ─ ─ ─ ─ ─┘
┌─ Regular expressions sublanguage ─ ─ ─ ─ ─ ─ ─ ─ ─ ─ ─ ─ ─ ─ ─ ─ ─ ─ ─ ─ ─ ─┐
│                                                                             │
│        $t \, @ \, \text{idle} \, s \rightarrow_c t \, @ \, s$            $t \, @ \, (\alpha;\beta) \, s \rightarrow_c t \, @ \, \alpha\beta \, s$          │
│                                                                             │
│        $t \, @ \, (\alpha|\beta) \, s \rightarrow_c t \, @ \, \alpha \, s$         $t \, @ \, (\alpha|\beta) \, s \rightarrow_c t \, @ \, \beta \, s$          │
│                                                                             │
│        $t \, @ \, \alpha* \, s \rightarrow_c t \, @ \, s$               $t \, @ \, \alpha* \, s \rightarrow_c t \, @ \, \alpha\alpha* \, s$            │
│                                                                             │
└─ ─ ─ ─ ─ ─ ─ ─ ─ ─ ─ ─ ─ ─ ─ ─ ─ ─ ─ ─ ─ ─ ─ ─ ─ ─ ─ ─ ─ ─ ─ ─ ─ ─ ─ ─ ─ ─ ─┘
┌─ Conditional ─ ─ ─ ─ ─ ─ ─ ─ ─ ─ ─ ─ ─ ─ ─ ─ ─ ─ ─ ─ ─ ─ ─ ─ ─ ─ ─ ─ ─ ─ ─ ─┐
│                                                                             │
│              $t \, @ \, (\alpha ? \beta : \gamma) \, s \rightarrow_c t \, @ \, \alpha\beta \, s$                    │
│                                                                             │
│         $t \, @ \, (\alpha ? \beta : \gamma) \, s \rightarrow_c t \, @ \, \gamma \, s$ if $t \, @ \, \alpha\theta \not\rightarrow^*_{s,c} t' \, @ \, \varepsilon$      │
│                                                                             │
└─ ─ ─ ─ ─ ─ ─ ─ ─ ─ ─ ─ ─ ─ ─ ─ ─ ─ ─ ─ ─ ─ ─ ─ ─ ─ ─ ─ ─ ─ ─ ─ ─ ─ ─ ─ ─ ─ ─┘
┌─ Subterm rewriting ─ ─ ─ ─ ─ ─ ─ ─ ─ ─ ─ ─ ─ ─ ─ ─ ─ ─ ─ ─ ─ ─ ─ ─ ─ ─ ─ ─ ─┐
│                                                                             │
│   $t \, @ \, (\text{matchrew } P(x_1,\ldots,x_n) \text{ s.t. } C \text{ by } x_1 \text{ using } \alpha_1, \ldots, x_n \text{ using } \alpha_n) \, s$ │
│           $\rightarrow_c \text{subterm}(\sigma_{-\{x_1,\ldots,x_n\}}(t); x_1 : \alpha_1\sigma, \ldots, x_n : \alpha_n\sigma) \, @ \, s$       │
│                                                                             │
│                   $q_k \rightarrow_\bullet q'_k$                                      │
│       ─────────────────────────────────────────────────────   $\bullet = s, c$  │
│       $\text{subterm}(t; \ldots, x_k : q_k, \ldots) \, @ \, s \rightarrow_\bullet \text{subterm}(t; \ldots, x_k : q'_k, \ldots) \, @ \, s$     │
│                                                                             │
└─ ─ ─ ─ ─ ─ ─ ─ ─ ─ ─ ─ ─ ─ ─ ─ ─ ─ ─ ─ ─ ─ ─ ─ ─ ─ ─ ─ ─ ─ ─ ─ ─ ─ ─ ─ ─ ─ ─┘
┌─ Strategy calls ─ ─ ─ ─ ─ ─ ─ ─ ─ ─ ─ ─ ─ ─ ─ ─ ─ ─ ─ ─ ─ ─ ─ ─ ─ ─ ─ ─ ─ ─ ─┐
│                                                                             │
│        $t \, @ \, sl(t_1, \ldots, t_n) \, s \rightarrow_c \delta\sigma s$ for any matching definition $(\delta, \sigma)$       │
│                                                                             │
└─ ─ ─ ─ ─ ─ ─ ─ ─ ─ ─ ─ ─ ─ ─ ─ ─ ─ ─ ─ ─ ─ ─ ─ ─ ─ ─ ─ ─ ─ ─ ─ ─ ─ ─ ─ ─ ─ ─┘
```

Fig. 1. Small-step operational semantics of the strategy language

Definition 2. Given $\pi \in \mathfrak{XS}^\omega$, we say that π iterates forever if any of the following conditions hold:

1. $\pi_k = t_k \, @ \, c_k \, \alpha* \, s$ for all $k \in \mathbb{N}$, and there are infinitely many $k \in \mathbb{N}$ such that $c_k = \varepsilon$ and $c_{k+1} = \alpha$.
2. $\pi_k = \text{subterm}(x_1 : \rho_{1,k}, \ldots, x_n : \rho_{n,k}; t)$ for all $k \in \mathbb{N}$, and $\rho_m = (\rho_{m,k})_{k \in \mathbb{N}}$ iterates forever for some $1 \leq m \leq n$.
3. π^k iterates forever for some $k \in \mathbb{N}$.

An execution π in $(\mathfrak{XS}, \rightarrow_{s,c})$ iterates finitely if it does not iterate forever, and an execution π in $(\mathfrak{XS}, \twoheadrightarrow)$ iterates finitely if its expansion to $\rightarrow_{s,c}$ transitions iterates finitely.

In simpler words, an execution iterates forever if the transition $t \, @ \, \alpha* \, s \rightarrow_c t \, @ \, \alpha\alpha* \, s$ is repeated infinitely many times for the same iteration, either inside a subterm state or at the top level. Definition 3 introduces the extensional strategy $E_K(\alpha, t) \subseteq E(\alpha, t)$ for a strategy expression α that respects the semantics of the Kleene star.

Definition 3. Given a strategy expression α and a term t,

$$E_K(\alpha, t) := \operatorname{cterm}(\operatorname{Ex}_K^\omega(\alpha, t))$$

where $\operatorname{Ex}_K^\omega(\alpha, t) := \{\pi \in \operatorname{Ex}^\omega(\alpha, t) \mid \pi \text{ iterates finitely}\}$.

The set $E_K(\alpha, t)$ is not necessarily closed, so in general it cannot be represented by an intensional strategy. While the Maude strategy language is Turing-complete under the original semantics [31, Proposition 3], the number of execution states is only finite if $E(\alpha, t)$ is a *closed* ω-regular language [31, Proposition 4]. With the new semantics we can drop the *closed* adjective from the theorem and represent any ω-regular language of executions with a finite number of execution states.

Proposition 1. *1. If the reachable states from $t @ \alpha$ by $\rightarrow_{s,c}$ are finitely many, then $E_K(\alpha, t)$ is an ω-regular language.*
2. If L is an ω-regular language, then there is a strategy expression β such that $\bigcup_{t \in T_\Sigma} E_K(\beta, t) = L$ and the reachable states from $t @ \beta$ by $\rightarrow_{s,c}$ are finitely many for all $t \in T_\Sigma$.

Proof. [28, Propositions 3.7 and 3.8]. □

However, this strategy cannot be represented in a plain transition system, because their executions are always closed. In the next section, we discuss how strategies interpreted under this semantics can be model checked.

5 Model Checking Fair Strategy-Controlled Models

Standard model-checking algorithms use plain transition systems as models, so they cannot directly incorporate other restrictions like those imposed by acceptance conditions on ω-automata. However, the automata-based LTL model-checking algorithm can handle this kind of restrictions with small changes, since the model is finally represented as an ω-automaton. Hence, we will represent the strategy-controlled Maude model directly as an ω-automaton instead of a Kripke structure.

At first sight, imposing that $t @ \alpha^* s \rightarrow_c t @ \alpha \alpha^* s$ transitions appear finitely often would be fine, since all iterations will be executed only finitely many times. However, this is too restrictive because of nonterminating recursive calls, which may start an iteration infinitely many non-consecutive times. For example, a recursive strategy mt like mt := rl1 * ; rl2 ; mt must be able to execute infinitely many times rl1 but only finitely many times between each pair of applications of rl2, i.e., in each strategy call. This complicates the definition of the acceptance conditions, and forces us to consider each distinct iteration separately. In order to identify a strategy and distinguish it from others, we introduce the following technical definitions.

Definition 4. For any $\pi \in \mathfrak{XS}^\omega$ and set of variables X, we define the following:

- A *position* in an execution state is a word $p \in X^*$. The substate of q at position p is written $q|_p$ and defined by $q|_\varepsilon = q$ and $(\text{subterm}(\ldots, x : q, \ldots ; t) @ s)|_{xp} = q|_p$.
- A *partial context* is a word $c \in (\text{Strat} \cup (X \to T_\Sigma) \cup X)^*$, and it is contained in an execution state $c \in q$ if $q = t @ c$ for some term t, or $c = c_1 x c_2$ for $x \in X$, $q = \text{subterm}(\ldots, x : q', \ldots ; t) @ c_2$ and $c_1 \in q'$.
- The position of a partial context $\text{pos}(c)$ is defined as $\text{pos}(\alpha s) = \text{pos}(s)$, $\text{pos}(\sigma s) = \text{pos}(s)$, $\text{pos}(xs) = x \, \text{pos}(s)$, and $\text{pos}(\varepsilon) = \varepsilon$.

The free variables in these sets are quantified existentially.

An iteration will be identified by its partial context $\alpha * c$. Two iterations are considered the same if their partial contexts coincide, and they always take place in fixed positions of the execution state. For tracking when a given iteration is started and finished during an execution, we extend the small-step operational semantics by tagging each control transition $\to_c^A$ with the set of iterations A in which it enters or leaves. For most axioms, we simply change $q \to_c q'$ to $q \to_c^\emptyset q'$, except for

$$t @ \alpha * s \to_c^{\{\text{enter}(\alpha * s)\}} t @ \alpha \, \alpha * s \qquad\qquad t @ \alpha * s \to_c^{\{\text{leave}(\alpha * s)\}} t @ s$$

and

$$\frac{q_k \to_c^A q_k'}{\text{subterm}(t; \ldots, x_k : q_k, \ldots) @ s \to_c^{\{cx_k s \mid c \in A\}} \text{subterm}(t; \ldots, x_k : q_k', \ldots) @ s}$$

Despite the tag being a set, a single control step can only enter or leave one iteration, so they are empty or singleton. For the $\twoheadrightarrow$ transition, we write $q \twoheadrightarrow^{A_1 \cup \ldots \cup A_n} q'$ if it expands to $q \to_c^{A_1} q_1 \to_c^{A_2} \cdots \to_c^{A_n} q_n \to_s q'$. The following result follows by definition of iterating finitely.

Proposition 2. *Given an execution π of the semantics, π iterates finitely iff for every partial context $\alpha^* c \in \pi$, $\text{enter}(\alpha * c)$ appears finitely often or $\text{leave}(\alpha * c)$ appears infinitely often in the transitions of π.*

Proof sketch. By induction, using that the conditions on enter and leave break item (1) of Definition 2. See [28, Proposition 4.8] for a similar proof.

The previous enter and leave tags can be used to define the acceptance conditions of the Streett automaton (see Sect. 2). Consider the graph of the semantics from an execution state $t @ \alpha$; assuming it is finite, finitely many iterations $C = \{\alpha_1 * c_1, \ldots, \alpha_n * c_n\}$ must have been executed to generate it. We can define the transition-based Streett automaton $(\mathfrak{XS}, T_{\Sigma/E}, Acc, \delta, \{t @ \alpha\}, F)$ where

$$Acc = \{\text{enter}(\alpha_k * c_k), \text{leave}(\alpha_k * c_k) \mid 1 \leq k \leq n\}$$
$$F = \{(\text{enter}(\alpha_k * c_k), \text{leave}(\alpha_k * c_k)) \mid 1 \leq k \leq n\}$$
$$\delta(q, t) = \begin{cases} \{(q', A) : q \twoheadrightarrow^A q'\} & \text{if cterm}(q) = t \\ \emptyset & \text{otherwise} \end{cases}$$

In summary, the Streett automaton is the tagged graph $(\mathfrak{XS}, \twoheadrightarrow)$ of the semantics with the acceptance conditions of Proposition 2.

Proposition 3. *Given a strategy α and an initial term t, the Streett automaton defined above accepts the language $E_K(\alpha, t)$.*

Proof sketch. By applying Proposition 2 on the runs of the Streett automaton.

For model checking, as usual, we also need to adapt the alphabet of the model to that of the propositional traces $\mathcal{P}(AP)$ in order to compute the product with the negated LTL property. However, it is enough to define the transition function as $\delta'(q, P) = \{\delta(q, t) \mid t \in T_{\Sigma/E}, \ell(t) = P\}$ for any $P \subseteq \mathcal{P}(AP)$.

Streett conditions can be translated into Bchi conditions by paying a blowup of $n(1 + k2^k)$ in the worst case where n is the number of states of the original automaton, and k the number of Streett pairs [10, Sect. 4.3.1]. Moreover, since the system automaton does no longer have trivial acceptance conditions, intersecting it with the property automaton is harder than in the usual case, and the number of states of the product automaton may double. Hence, considering the iteration under the new semantics is more costly, and simplifying and reducing the number of Streett pairs is always convenient. One possibility is to ignore iterations that do not produce cycles, and another one is merging Streett pairs when possible. For example, since iterations can only be repeated infinitely often because of tail recursive calls, we can define $t @ sl(t_1, \ldots, t_n)\, s \to_c^{\{\mathsf{call}(\sigma s)\}} \delta\sigma s$ and remove the leave tags. An iteration $\alpha^* s' \sigma s$ where s' does not contain other call contexts would be accepted if $\mathsf{enter}(\alpha^* s' \sigma s)$ appears finitely often or $\mathsf{call}(\sigma s)$ appears infinitely often. This allows merging Streett pairs of iterations that belong to the same recursive call if s' does not contain a variable (i.e. if they are not in different substates of a subterm rewriting operator).

5.1 Implementation

The model-checking procedure described in this paper is implemented in Python as part of the umaudemc tool [29,30]. Spot [15] is used for building the two ω-automata and checking the emptiness of their product. Since the strategy-controlled rewrite graph provided by the maude Python library does not trace the execution of iterations, we generate the semantics graph with an extension of a built-in implementation of the strategy language included in umaudemc for a previous work [27].[3] This model is then scanned to accumulate the iterations and identify the Streett pairs. Using the Python API of Spot, a transition-based Streett automaton is built, and the emptiness of its product with another automaton for the negated LTL formula is checked. In case the property is not satisfied, this yields a counterexample run that is presented to the user. This new model checker is available through the check subcommand of the umaudemc

---

[3] In a previous prototype [28], an executable Maude specification of the small-step operational semantics in Sect. 3 was used, and it is still used for the other logics.

```
            cup(ball)  cup(nothing)  nothing
                 uncover│leave(α*c)    cover  ∅
            cup(ball)  cup(nothing)  cup(nothing)
                    swap│enter(α*c)
            cup(nothing) cup(nothing)  cup(ball)
```

Fig. 2. Small subgraph of the Kleene-aware model of the running example

interface, where the flag --kleene-iteration (abbreviated to -k) enables this interpretation.

Back to the running example, the property $\Box \Diamond$ uncovered should be satisfied under the cups strategy when the iteration is understood as the Kleene star. The check command shows the expected result with the -k flag.

```
$ umaudemc check cupsballs.maude initial '[] <> uncovered' cups -k
The property is satisfied in the initial state
(16 system states, 102 rewrites, 2 Bchi states)
```

In this case, a single distinct iteration has been found, whose partial context is swap * uncover ; cover ; cups. The graph of the strategy-controlled model with the iteration flags can be obtained using the graph command of umaudemc. A small subgraph of the graph generated by the following command is shown in Fig. 2.

```
$ umaudemc graph cupsballs.maude initial cups -k
```

The umaudemc tool supports model checking under the standard semantics of the Maude strategy language for other logics like CTL and CTL*, using external tools as backends. This is also possible for this new semantics using an alternative but related approach.

5.2 An Alternative Approach for LTL and Other Logics

Instead of adapting the guts of the model-checking algorithm, we can consider an alternative approach that discharges the fairness restrictions to the property logic. In order to do that, we need to push the enter and leave annotations from the edges to the states as artificial atomic propositions. The number of model states may increase as a result, since the same original state may be reached with or without using an iteration. The following proposition proves that this is generally possible with LTL, and the procedure can be similarly applied to other logics.

Proposition 4. *Given a Maude strategy expression α, an LTL property φ, and assuming that the iterations that occur in the reachable states from $t @ \alpha$ are numbered from 1 to n, $\pi \vDash \varphi$ for all $\pi \in E_K(\alpha, t)$ iff $\mathcal{M}_{\alpha,t}^K \vDash \psi \to \varphi$ where*

$$\mathcal{M}_{\alpha,t}^K := (\mathcal{XS} \times \mathcal{P}(AP'), \twoheadrightarrow', \{t @ \alpha\}, AP \cup AP', \ell')$$

with $AP' = \{e_1, l_1, \ldots, e_n, l_n\}$, $e_k = \text{enter}(\alpha_k * c_k)$, $l_k = \text{leave}(\alpha_k * c_k)$, $\ell'((q, A)) = \ell(q) \cup A$, $(q, A) \rightarrow' (q', A')$ if $q \rightarrow^{A'} q'$, $\psi = \bigwedge_{k=1}^{n} \psi_k$, and $\psi_k = \Diamond \Box \neg e_k \vee \Box \Diamond l_k$.

Using the Maude LTL model checker, the previous property is verified by the second method, which will actually check $(\Diamond \Box \neg e_1 \vee \Box \Diamond l_1) \rightarrow \Box \Diamond$ uncovered.

```
$ umaudemc check cups.maude initial '[] <> uncovered'
           cups -k --backend maude
The property is satisfied in the initial state
(27 system states, 6384 rewrites, 7 Bchi states)
```

We have passed --backend maude, instead of the implicit --backend spot when -k is present, to use the built-in Maude model checker (and so the alternative approach) instead of Spot.

The finite-iteration semantics can also be respected in CTL* by adding the restriction ψ as premise to all path quantifiers, i.e. replacing ϕ in each $\mathbf{A} \phi$ by $\psi \rightarrow \phi$ and in each $\mathbf{E} \phi$ by $\psi \wedge \phi$. In particular, CTL properties can be transformed likewise yielding CTL* properties, and the same effect can be achieved with the fairness constraints supported by some model checkers [10]. However, the complexity of the transformed formulae makes it impractical for large examples. For example, the property $\mathbf{A} \Box \mathbf{E} \Diamond \text{hit}$ says that it is always possible to get the guess right, and it is translated into $\mathbf{A} (\Diamond \Box \neg e \vee \Box \Diamond l \rightarrow \Box (\mathbf{E}((\Diamond \Box \neg e \vee \Box \Diamond l) \wedge \Diamond \text{hit})))$ for the following problem.

```
$ umaudemc check cupsballs.maude initial
       'A [] E <> hit' cups -k --backend pymc
The property is satisfied in the initial state
(27 system states, 6846 rewrites, holds in 27/27 states)
```

While previous executions were instantaneous, this one takes around 4.5 min using pyModelChecking [8] as backend.

Finally, illusionists and conmen sometimes make the ball disappear to obtain the admiration or the money of the public. Adding this possibility to our rules and strategies is a small change.

```
    rl [disappear] : cup(ball)    => cup(nothing) .
    rl [appear]    : cup(nothing) => cup(ball)    .

    sd cups2 := (swap | disappear) * ;
                (amatch cup(call) ? idle : fail) ;
                uncover ; cover ; appear ; cups2 .
```

According to the cups2 definition, the ball can be distracted at any time while swapping the cups with the disappear rule, and the test ensures that this is actually done. Then the property $\mathbf{A} \Box \mathbf{E} \Diamond \text{hit}$ is no longer satisfied.

```
$ umaudemc check cupsballs.maude initial
    'A [] E <> hit' cups2 -k --backend pymc
The property is not satisfied in the initial state
(15 system states, 9739 rewrites, holds in 0/15 states)
```

Another application of this model checker is shown in [28, Sect. 6.3], where the meaning of the iteration ensures that some processes in a processor release it infinitely often.

6 Related Work

The Kleene star semantics for the Maude strategy language presented in this paper is based on the small-step operational semantics defined in [31]. In that work, we also study the expressivity of the Maude strategy language and identify the limitations solved in this paper, and introduce along with [30] the model-checking approach for strategy-controlled systems that we have taken as a reference here. Indeed, the implementation of the model checker is integrated into the umaudemc tool [29] presented in [30].

In the literature, there are several works dedicated to model checking with fairness constraints [10], even in the context of Maude [2], and to fair models of computation like fair Kripke structures [20]. NuSMV [9] is a well-known example of CTL model checker that supports fairness constraints as separate input. In a wider sense, we are considering constraints involving unquantified nondeterminism, which has always been a relevant topic in the study of concurrency [12, §9].

We also want to mention Meseguer's Temporal Logic of Rewriting (TLR) [24], which is introduced along with a strategy language to be used as a dynamic logic for checking system properties. This language shares most of their operators with the final Maude strategy language, and namely it includes an iteration operator α+, which is available in the Maude strategy language as a synonym of α ; α*. However, the semantics of that strategy language is defined for finite computations, so the subtlety of the Kleene star is not present.

7 Conclusions

We have proposed a different semantics of the Maude strategy language that solves the inaccuracy of the Kleene star operator and allows specifying constraints on the whole-run behavior of the model, like fairness, as part of the strategy. Indeed, any ω-regular sublanguage of the executions of the model can be described with this version of the strategy language, while keeping the number of execution states finite. Maude specifications controlled by such a strategy can be model checked using an extension of the umaudemc tool, which relies on the Spot ω-automata library. However, we have not set it as the default semantics for the model checker because the verification algorithms are more expensive, and cannot be applied on the fly (i.e. without generating the full model in advance).

CTL and CTL* properties can also be checked with this tool by translating the input formula and adding artificial atomic propositions to delimit iterations.

Acknowledgments. Research partially supported by the Spanish AEI through project ProCode (PID2019-108528RB-C22/AEI/10.13039/501100011033).

References

1. Bae, K., Escobar, S., Meseguer, J.: Abstract logical model checking of infinite-state systems using narrowing. In: van Raamsdonk, F. (ed.) RTA 2013. LIPIcs, vol. 21, pp. 81–96. Schloss Dagstuhl - Leibniz-Zentrum für Informatik (2013). https://doi.org/10.4230/LIPICS.RTA.2013.81
2. Bae, K., Meseguer, J.: Model checking linear temporal logic of rewriting formulas under localized fairness. Sci. Comput. Program. **99**, 193–234 (2015). https://doi.org/10.1016/j.scico.2014.02.006
3. Balland, E., Brauner, P., Kopetz, R., Moreau, P., Reilles, A.: Tom: Piggybacking rewriting on Java. In: Baader, F. (ed.) RTA 2007. LNCS, vol. 4533, pp. 36–47. Springer, Berlin (2007). https://doi.org/10.1007/978-3-540-73449-9_5
4. Borovanský, P., Kirchner, C., Kirchner, H., Ringeissen, C.: Rewriting with strategies in ELAN: A functional semantics. Int. J. Found. Comput. Sci. **12**(1), 69–95 (2001). https://doi.org/10.1142/S0129054101000412
5. Bouhoula, A., Jouannaud, J.P., Meseguer, J.: Specification and proof in membership equational logic. In: Bidoit, M., Dauchet, M. (eds.) TAPSOFT'97. LNCS, vol. 1214, pp. 67–92. Springer, Berlin (1997). https://doi.org/10.1007/BFb0030589
6. Bourdier, T., Cirstea, H., Dougherty, D.J., Kirchner, H.: Extensional and intensional strategies. In: Fernández, M. (ed.) WRS 2009. Electronic Proceedings in Theoretical Computer Science, vol. 15, pp. 1–19 (2009). https://doi.org/10.4204/EPTCS.15.1
7. Bravenboer, M., Kalleberg, K.T., Vermaas, R., Visser, E.: Stratego/XT 0.17. A language and toolset for program transformation. Sci. Comput. Program. **72**(1–2), 52–70 (2008). https://doi.org/10.1016/j.scico.2007.11.003
8. Casagrande, A.: pyModelChecking (2020). https://pypi.org/project/pyModelChecking
9. Cimatti, A., Clarke, E.M., Giunchiglia, E., Giunchiglia, F., Pistore, M., Roveri, M., Sebastiani, R., Tacchella, A.: NuSMV 2: An opensource tool for symbolic model checking. In: Brinksma, E., Larsen, K.G. (eds.) CAV 2002. LNCS, vol. 2404, pp. 359–364. Springer, Berlin (2002). https://doi.org/10.1007/3-540-45657-0_29
10. Clarke, E.M., Henzinger, T.A., Veith, H., Bloem, R. (eds.): Handbook of Model Checking. Springer, Berlin (2018). https://doi.org/10.1007/978-3-319-10575-8
11. Clavel, M., Durán, F., Eker, S., Escobar, S., Lincoln, P., Martí-Oliet, N., Meseguer, J., Rubio, R., Talcott, C.: Maude Manual v3.4 (2024). https://maude.lcc.uma.es/maude-manual
12. Dijkstra, E.W.: A Discipline of Programming. Prentice-Hall (1976). https://www.worldcat.org/oclc/01958445

13. Durán, F., Eker, S., Escobar, S., Martí-Oliet, N., Meseguer, J., Rubio, R., Talcott, C.L.: Programming and symbolic computation in Maude. J. Log. Algebraic Methods Program. **110** (2020). https://doi.org/10.1016/J.JLAMP.2019.100497
14. Duret-Lutz, A., Poitrenaud, D.: SPOT: an extensible model checking library using transition-based generalized Büchi automata. In: DeGroot, D., Harrison, P.G., Wijshoff, H.A.G., Segall, Z. (eds.) MASCOTS 2004, pp. 76–83. IEEE Computer Society (2004). https://doi.org/10.1109/MASCOT.2004.1348184
15. Duret-Lutz, A., Renault, E., Colange, M., Renkin, F., Aisse, A.G., Schlehuber-Caissier, P., Medioni, T., Martin, A., Dubois, J., Gillard, C., Lauko, H.: From Spot 2.0 to Spot 2.10: What's new? In: Shoham, S., Vizel, Y. (eds.) CAV 2022, Part II. LNCS, vol. 13372, pp. 174–187. Springer, Berlin (2022). https://doi.org/10.1007/978-3-031-13188-2_9
16. Eker, S., Martí-Oliet, N., Meseguer, J., Rubio, R., Verdejo, A.: The Maude strategy language. J. Log. Algebraic Methods Program. **134**, 100887 (2023). https://doi.org/10.1016/J.JLAMP.2023.100887
17. Eker, S., Meseguer, J., Sridharanarayanan, A.: The Maude LTL model checker. In: Gadducci, F., Montanari, U. (eds.) WRLA 2002. ENTCS, vol. 71, pp. 162–187. Elsevier (2004). https://doi.org/10.1016/S1571-0661(05)82534-4
18. Fernández, M., Kirchner, H., Pinaud, B.: Strategic port graph rewriting: an interactive modelling framework. Math. Struct. Comput. Sci. **29**(5), 615–662 (2019). https://doi.org/10.1017/S0960129518000270
19. Gastin, P., Oddoux, D.: Fast LTL to Büchi automata translation. In: Berry, G., Comon, H., Finkel, A. (eds.) CAV 2001. LNCS, vol. 2102, pp. 53–65. Springer, Berlin (2001). https://doi.org/10.1007/3-540-44585-4_6
20. Kesten, Y., Pnueli, A., Raviv, L.: Algorithmic verification of linear temporal logic specifications. In: Larsen, K.G., Skyum, S., Winskel, G. (eds.) ICALP'98. LNCS, vol. 1443, pp. 1–16. Springer, Berlin (1998). https://doi.org/10.1007/BFB0055036
21. Marin, M., Kutsia, T.: Foundations of the rule-based system ρlog. J. Appl. Non Class. Logics **16**(1–2), 151–168 (2006). https://doi.org/10.3166/jancl.16.151-168
22. Martí-Oliet, N., Palomino, M., Verdejo, A.: Rewriting logic bibliography by topic: 1990–2011. J. Log. Algebr. Methods Program. **81**(7–8), 782–815 (2012). https://doi.org/10.1016/j.jlap.2012.06.001
23. Meseguer, J.: Conditional rewriting logic as a unified model of concurrency. Theor. Comput. Sci. **96**(1), 73–155 (1992). https://doi.org/10.1016/0304-3975(92)90182-F
24. Meseguer, J.: The temporal logic of rewriting. Technical Report UIUCDCS-R-2007-2815, Department of Computer Science, University of Illinois at Urbana-Champaign (2007). http://hdl.handle.net/2142/11293
25. Pnueli, A.: The temporal logic of programs. In: FOCS 1977, pp. 46–57. IEEE Computer Society (1977). https://doi.org/10.1109/SFCS.1977.32
26. Rubio, R.: Maude as a library: An efficient all-purpose programming interface. In: Bae, K. (ed.) WRLA 2022. LNCS, vol. 13252, pp. 274–294. Springer, Berlin (2022). https://doi.org/10.1007/978-3-031-12441-9_14
27. Rubio, R., Martí-Oliet, N., Pita, I., Verdejo, A.: QMaude: Quantitative specification and verification in rewriting logic. In: Chechik, M., Katoen, J., Leucker, M. (eds.) FM 2023. LNCS, vol. 14000, pp. 240–259. Springer, Berlin (2023). https://doi.org/10.1007/978-3-031-27481-7_15
28. Rubio, R.: Model checking of strategy-controlled systems in rewriting logic. Ph.D. thesis, Universidad Complutense de Madrid (2022). https://hdl.handle.net/20.500.14352/3553

29. Rubio, R.: Unified Maude model-checking tool (umaudemc) (2024). https://doi.org/10.5281/zenodo.7339535, https://github.com/fadoss/umaudemc
30. Rubio, R., Martí-Oliet, N., Pita, I., Verdejo, A.: Strategies, model checking and branching-time properties in Maude. J. Log. Algebr. Methods Program. **123** (2021). https://doi.org/10.1016/j.jlamp.2021.100700
31. Rubio, R., Martí-Oliet, N., Pita, I., Verdejo, A.: Model checking strategy-controlled systems in rewriting logic. Automat. Softw. Eng. **29**(1) (2022). https://doi.org/10.1007/s10515-021-00307-9

Tool Papers

The Hrewrite Library: A Term Rewriting Engine for Automatic Code Assembly

Michael Lienhardt(✉)

DTIS, ONERA, Université Paris-Saclay, 91120 Palaiseau, France
michael.lienhardt@onera.fr

Abstract. This paper presents the hrewrite library, that implements a term rewriting engine used as a backend for the generation of code assemblies. This library is implemented in C++ with a Python API, and accepts regular expression order-sorted term signatures and conditional rewriting rules. This paper describes the motivations of the library, its design, and illustrates it with some basic usage and with a discussion of its role as code assembly generation backend.

1 Introduction

The problem of automatically assembling code components to obtain a correct, complex and efficient behavior has received a lot of attention over the past decades, e.g., in [1,4,12,16,20]. In many of these works, the code components are packages and the necessity to assemble them comes from a name-based notion of dependency: each component C is tagged with a Boolean formula over component names that states which other components must be present (or not) for C to be functional. As discussed in [1,20], such problems can be managed by a SAT solver. In [12,16], the dependencies are modeled with numerical constraints, and can be managed by a minizinc or SMT solver.

Recently a new instance of the automatic code assembly problem was investigated in [21], where code components are functions with multiple inputs and outputs and these functions must be combined in such a way as to produce a value desired by the end user. Consequently, the notion of dependency in this new instance is based on the specification of values, and none of the approaches described previously can be reused. In [21], the authors describe a solution for this new instance, based on a new term rewriting engine: the current paper

---

This work was partially supported by the SONICE project, granted by the French Directorate General for Civil Aviation (DGAC).

complements [21] by presenting this new engine, called *hrewrite*[1]. This engine is implemented as a C++ pure-header library with a Python interface, and includes the following features:

(i) it implements rewriting over unranked trees described with regular expression order-sorted term signatures, as discussed in [18];
(ii) it implements conditional rewriting rules where guards are arbitrary Python functions;
(iii) it allows for terms to store any Python objects; and
(iv) it ensures that no term is duplicated in memory (consequently a term is a DAG in *hrewrite*).

The outline of the paper is as follows: Sect. 2 describes the motivations for the new rewriting engine and its features; Sect. 3 describes the overall architecture of the library; Sect. 4 presents the Python API of the library via a small representative set of examples; Sect. 5 briefly discusses the usage of the library as a backend for automatic code assembly; Sect. 6 discusses the related work; and Sect. 7 concludes the paper.

2 Motivations

Likewise to similar problems, to solve the dependency analysis problem raised in [21], we need to answer two main questions: how to specify the inputs and outputs of every function; and how to match them in order to build a correct assembly. We structure this section in three parts: (i) we discuss the unranked tree nature of the values exchanged between the functions; (ii) we discuss the need to abstract from these values to express the dependencies of some functions; and (iii) we briefly motivate why *hrewrite* generates a DAG and not a tree.

The Values. The application domain of the functions considered in [21] is *Computational Fluid Dynamics*, i.e., the functions compute physical values (like pressure or speed) of a fluid evolving in a 3D mesh. The main standard to describe such meshes and the value they carry is called CGNS [19], and like a DTD file for XML [8,25], it corresponds to an unranked tree language. Moreover, the CGNS standard splits a mesh in a graph of *zones* (that model the volume in which the fluid evolves) and *borders* (that model the borders of the mesh, e.g., the walls, or the places where the fluid is injected/extracted). This standard has thus three main consequences on how to specify functions.

1. Most functions are *generic* since they can compute a physical value on any zone of the mesh. This implies that our specifications must include a notion of variable.
2. While most values hosted on a zone can be described with a simple name (like "pressure"), some functions manipulate subtrees of a CGNS file. Consequently, our specifications must include a notion of unranked tree language.

---

[1] The *hrewrite* library is available at https://github.com/onera/hrewrite.

3. Moreover, CGNS trees can carry base values (like strings or numpy tensors) that are central for the configuration of the functions, before their execution. Hence, our specifications must be able to describe these values.

It is thus clear that: the unranked tree model as described in [18], extended with arbitrary Python values (to efficiently store CGNS-carried data), is a good fit to specify values in [21]; and pattern matching (extended with equality testing between Python values) is enough to perform the dependency analysis.

The Abstractions. To illustrate the kind of abstraction needed to express some of the functions' dependencies, let us consider the borders of a mesh. These borders carry physical properties to describe how they affect the fluid flow, and so many functions computing a value on a zone need to access the borders linked to that zone. As these functions are generic (they can be applied on any zones), we need a way to: (*i*) express the set of borders linked to a variable; and (*ii*) once the value of that variable is known, translate it into the actual set of borders.

For instance, if we consider the function convectiveFlux computing the convective flow, we need to be able to express its dependencies as follows (using the syntax described in [21]):

```
vars Z : zone
fun convectiveFluxBC:
  inputs
    data(Primitive, Z)
    data(Conservative, bordersOfZone(Z))
  outputs data(Fxc, Z)
```

In this specification, Z is a variable of zone; data is a constructor putting a value on a specific element of the mesh; Primitive and Conservative are values (which can exist on zones and on borders); bordersOfZone is a constructor corresponding to getting the list of borders of the zone in parameter; and Fxc is the convective flow computed by the operator. Note that following this specification, the number of inputs of the function convectiveFlux is at least one but has no upper bound: data(Primitive, bordersOfZone(Z)) corresponds to one input per border of Z, and the number of borders of a zone is unknown when specifying a function. Consequently, in addition to the mechanism to translate bordersOfZone(Z) into the list of borders of Z when it is known, we also need a mechanism to distribute the data constructor over that list.

To be able to easily express these mechanisms (and few others of similar nature), we extended the pattern matching previously discussed into rewriting, with conditional rewriting rules where the guards can be any user-defined function: to manage the bordersOfZone constructor, we automatically generate from a CGNS file the rewriting rules encoding its graph; to manage the distribution of the data constructor, we write few rewriting rules; and we use the user-defined guards to fetch useful information from the rest of the CFD program.

Shared Data. While a rewriting engine is useful to put a function's specification into normal form and to match inputs with outputs, it can also be used (as described in [21]) to perform the dependency analysis, thus generating the complete code assembly as a term. In practice, we observed that many intermediate values in the code assembly are used by multiple functions (3 in average), and so storing a term as a DAG instead of a tree avoids a lot of replication: we had an example of code assembly consisting of 384 nodes that would correspond to a tree of 150983896 nodes.

3 Architecture of the Library

The core of *hrewrite* is implemented as a pure C++17 header generic library, which is then instantiated and wrapped to provide the Python API. Figure 1 presents the overall architecture of *hrewrite*, which is structured in at least 6 modules implemented in C++, plus one Python module.

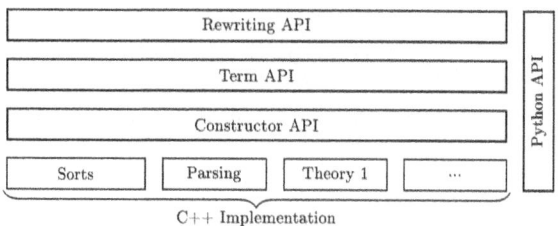

Fig. 1. Structure of the *hrewrite* library

The Sorts Module. This module provides the API to declare new sorts and the subsort relation. It is also responsible for storing this information and can be queried, for instance to ensure that the parameters of a term constructor are valid.

The Parsing Module. This module provides all the functionalities necessary to manage regular expressions. First, it translates regular expressions into objects (e.g., non-deterministic finite automata) capable to recognise words of the corresponding language. Note that the generated objects are linked to the *Sorts* module so the subsort relation is taken in account during world recognition. Moreover, these objects are used as basis for the implementation of the pattern matching which will be discussed a bit later.

The Theories Modules. Each of these modules correspond to a possible implementation of a term node, and can be freely combined to enable different functionalities in terms. For instance, in the core implementation of *hrewrite*, 4 theory modules have been implemented: one for variables; one for structured terms; one for terms with no subterms; and one templated module for leaf term nodes holding a value of a user-given type.

The Constructor API Module. This module provides the API to declare new constructors for any of the included theories (except for the variable theory). It is also responsible for storing these constructors and their signatures, which is used to ensure that a user-defined term is well constructed. This module does not fully implement the functionalities described in [18], as it allows for a term constructor to be declared at most once: i.e., a term constructor cannot be overloaded and can only have one sort.

The Term API Module. This module provides several functionalities related to terms. First, it has the API to construct new terms: either variables, or using the constructors declared in the *Constructor API* Module. It also provide all the base lookup functionalities on terms, like checking if a term is a variable, accessing the subterm of a structured term or accessing the value of a value-holding term. Moreover, this module implements the pattern matching on terms. This functionality is implemented with a simple backtracking algorithm, which is motivated by the fact that multiple matches are possible for a given pattern, and the user-defined guard may choose any of the possible matches. Currently, only linear patterns are supported. Finally, this module can also store the created terms to ensure that a term is created at most once in memory.

The Rewriting API Module. This module provides the API for registering a set of rewriting rules and applying it on a ground term. The rewriting rules are stored in a simple mapping that gives for every term constructor c the list of rewriting rules whose pattern has c as root constructor, which allows for relatively quick rewriting rule lookup. Both the innermost and outermost rewriting strategies have been implemented. The innermost strategy traverses the input term starting by its leaves, and for every encountered subterm t with root constructor c, it iterates over the rewriting rules in the c list to check if one can be applied. If several rewriting rules matches t, then the first one will be applied. The outermost strategy traverses the input term in both direction, first from its root to its leaves, and then back, and for each encountered subterm, it checks if it can be rewritten the same way as the innermost strategy. Finally, this module can also store a mapping giving the normal form of every rewritten term, which can be useful when the same term is rewritten several times.

The Python API Module. As previously described, this module first instantiates the generic functionalities of *hrewrite*, and wraps them in a user-friendly Python API. The instantiation includes all the theories implemented in the *hrewrite* library, with the templated theory for value instantiated for Python objects. Moreover, to ensure that our terms are DAGs with no possible duplication of nodes encoding a computation, the optional storage of the *Term API* module has been selected; we also selected the optional storage of the normal forms of terms to avoid rewriting multiple times the same term, which happens often in our usecase. Indeed, as discussed in Sect. 2 in our code assemblies, each data is used 3 times in average, and without the normal form storage, each of these usages triggers a new rewriting process.

4 The Python API

In this section, we present a simple example illustrating the Python API of *hrewrite*. To use our library, we first need to import it:

```
import hrewrite as hrw
```

Our example is structured as follows: we first construct a simple algebra for natural numbers; then we illustrate the value-carrying terms by rewriting terms into Python values; and conclude with a simple algebra for lists.

Algebra for Natural Numbers. We start by declaring three sorts (one for the zero value, one for the non-zero values, and one for all natural integers) with the corresponding subsort relation:

```
hrw.sorts("Zero", "NzNat", "Nat")
hrw.subsort("Zero", "Nat")
hrw.subsort("NzNat", "Nat")
```

Next, we declare the following term constructors:

```
zero = hrw.constructor("zero", hrw.free() >> "Zero")
s = hrw.constructor("s", hrw.free("Nat") >> "NzNat")
```

Line 1 declares the constructor named "zero", that takes no parameters and is of the sort "Zero", and stores it in the variable zero[2]. Line 2 declares the constructor named "s", that takes one parameter of sort "Nat" and returns a term of sort "NzNat", and stores it in the variable s. Terms are constructed and printed as follows:

```
two = s(s(zero))
print(two)
```

Finally, let us declare the plus function and its semantics as rewriting rules:

```
plus = hrw.constructor("plus", hrw.free("Nat Nat") >> "Nat")
alpha, beta = hrw.vars("Nat", "Nat")
rw_eng = hrw.rw_engine_cls()
rw_eng.add(plus(zero, alpha), alpha)
rw_eng.add(plus(s(alpha), beta), plus(alpha, s(beta)))
```

Line 1 declares the constructor named "plus", that takes two parameters of sort "Nat" and returns a term of sort "Nat", and stores it in the variable plus. Line 2 declares the variables alpha and beta, both of sort "Nat". Line 3 declares a new rewriting context rw_eng, and Lines 4 and 5 introduce the standard rewriting rules for the semantics of plus. Rewriting the term 2+3 is done as follows:

```
five = rw_eng.rewrite(plus(two, s(two)))
```

Natural Numbers to Python. To store Python number in our terms, we first create a new sort (with the corresponding subsort relation) and a new constructor for this container:

```
hrw.sort("Val")
hrw.subsort("Val", "Nat")
val = hrw.constructor("val", hrw.lit() >> "Val")
```

In line 3, hrw.lit() states that the term constructor val holds a Python value. We model the conversion from pure term naturals to Python naturals as follows:

---

[2] The name of the hrw.free function comes from the *free* theory in Maude, that implements basic structured terms

```
rw_eng.add(zero, val(0))
valpha, vbeta = hrw.vars("Val", "Val")
def guard_s(rw_eng, substitution):
   t = hrw.instantiate(valpha, substitution)
   substitution.add(vbeta, val(hrw.get_value(t) + 1))
   return True
rw_eng.add(s(valpha), vbeta, guard_s)
```

Line 1 simply states that the term zero corresponds to 0. Lines 2 to 7 encode the semantics of s: the increment of the integer value is performed within the guard guard_s. Line 7 states that the successor of a Python value is rewritten to a variable vbeta, whose image in the matching substitution is set in guard_s. A guard (in line 3) is a Python function that takes two parameters: the rewriting engine executing the guard (in case some rewriting must be performed in the guard), and the substitution computed by the pattern matching. In line 4, the guard extracts the image t of valpha from the substitution, and in line 5, it sets the value of vbeta to be a val containing the value contained in t plus 1. Finally, in line 6, the function returns True to signal the rewriting engine that the guard has been validated.

Lists of Natural Numbers. Since *hrewrite* manages unranked trees, the constructor for list of natural numbers can be declared as follows:

```
hrw.sorts("list")
List = hrw.constructor("List", hrw.free("Nat*") >> "list")
```

As indicated by the star in "Nat*" in line 2, the List constructor can take any number of natural numbers in parameter. The concatenation of two lists is also relatively simple to express:

```
concat = hrw.constructor(
   "concat", hrw.free("list list") >> "list")
vl1, vl2 = hrw.vars("Nat*", "Nat*")
rw_eng.add(concat(List(vl1), List(vl2)), List(vl1, vl2))
```

The rewriting rule in line 3 simply puts the content of the first list in vl1, the content of the second list in vl2, and then puts these two contents in sequence in the resulting list.

5 Case Study

In [21], *hrewrite* is used to automatically generate 597 code assemblies representative of computations in CFD. In this section, we first provide a simplified example illustrating this automatic generation, and then gives an initial benchmark on the speed of the *hrewrite* library.

5.1 Example of Code Assembly

For simplicity, we will not present the specificities of CGNS trees, and use a Maude-like DSL to illustrate our signature and rewriting rules. Our example is structured as follows: (*i*) we introduce the core signature that models data and function applications, together with few rewriting rules that, as discussed in Sect. 2, encode the distribution of `data` over lists of borders; (*ii*) we give the specification of several functions that can be used in our assembly; (*iii*) we then present the rewriting rules that are automatically generated from this specification and from an example mesh; and (*iv*) we apply the rewriting rules on a term corresponding to a data that must be computed, and present the resulting DAG.

```
 1  sort field zone border borderType .         10  op dataList(_): data* -> data .
 2  sort value data functionID appStatus .      11  op fieldList(_): field* -> field .
 3  subsort zone < field .                      12  op zoneOfBorder(_): border -> field .
 4  subsort border < field .                    13  op bordersOfZone(_): zone -> field .
 5                                              14  op getBorderType(_): border -> borderType .
 6  op data(_,_): value field -> data .         15  op wall: -> borderType .
 7  op appNone(_): data -> appStatus .          16  op inputFlow: -> borderType .
 8  op appDone(_,_,_):                          17  op outputFlow: -> borderType .
 9     data functionID appStatus* -> appStatus .
18  vars V: value, F: field, Fs: field*, D: data, Ds: data* .
19
20  rl data(V, fieldList(F, Fs)) => dataList(data(V, F), data(V, fieldList(Fs))) .
21  rl data(V, fieldList()) => dataList() .
22  rl appNone(dataList(D, Ds)) => appList(appNone(D), appNone(dataList(Ds))) .
23  rl appNone(dataList()) => appList() .
```

Fig. 2. A simple signature used for automatic function assembly

Signature and Distribution. Figure 2 first presents the signature structuring our function assembly (Lines 1–17) and the rewriting rules for distribution (Lines 18–23). The `field` sort corresponds to elements of a mesh, which can be either a `zone` (corresponding to a finite volume) or a `border` (corresponding to a border of a zone). A border can be of several types. An example of such a mesh is presented in Fig. 3, that contains only one zone, with four borders. Two of these borders are walls blocking the flow of the fluid; the border on the left injects more fluid in the zone; and the border on the right allows for the fluid to escape. The data manipulated by our functions are values hosted on the different elements of the mesh: this is modeled with the `data` constructor that takes in parameter a value and a field. The `appNone` constructor states that the data in parameter is not computed by any function, while the `appDone` constructor states that the data is computed by a function (of sort `FunctionID`) applied on the parameters in the `appStatus` list.

Moreover, we have a container for lists of data and another one for lists of fields; `zoneOfBorder` models the function getting the zone attached to a border; `bordersOfZone` models the function getting the list of borders attached to a zone; `getBorderType` models the function getting the type of a border; and `wall`, `inputFlow` and `outputFlow` are the three border types considered in our example.

Finally, our rewriting rules encodes two distributions: one for the `data` constructor over its `field` parameter, and another one for the `appNone` constructor over its `data` parameter.

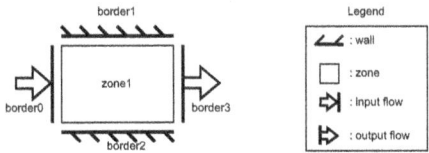

Fig. 3. A Mesh example, with one zone and four borders

```
 1  vars Z: zone, B: border                    18  fun explicitIncrement:
 2                                             19    inputs data(Balance, Z)
 3  fun primitive:                             20    outputs data(Rhs, Z)
 4    inputs data(Conservative, Z)             21
 5    outputs data(Primitive, Z)               22  fun inlet:
 6  fun convectiveFlux:                        23    inputs data(Conservative, zoneOfBorder(B))
 7    inputs data(Primitive, Z)                24    outputs data(Conservative, B)
 8    outputs data(Fxc, Z)                     25    guard bc_type(B) == inputFlow
 9  fun convectiveFluxBC:                      26  fun outpres:
10    inputs                                   27    inputs data(Conservative, zoneOfBorder(B))
11      data(Primitive, Z)                     28    outputs data(Conservative, B)
12      data(Conservative, bordersOfZone(Z))   29    guard bc_type(B) == outputFlow
13    outputs data(FxcBC, Z)                   30  fun wall:
14                                             31    inputs data(Conservative, zoneOfBorder(B))
15  fun fluxBalance:                           32    outputs data(Conservative, B)
16    inputs data(Fxc, Z) data(FxcBC, Z)       33    guard bc_type(B) == wall
17    outputs data(Balance, Z)
```

Fig. 4. Function specifications

Function Specifications. Figure 4 specifies the signature (using a DSL similar to [21]) of the different functions that can be used to construct our assembly. Inputs and outputs are specified with terms[3] that can contain variables: here, the variables Z of sort zone and B of sort border are used. The primitive function takes in parameter the Conservative value on a zone (which models the currently computed flow on the zone), and normalizes it into a value called Primitive. The convectiveFlux function takes in parameter the Primitive value on a zone, and computes Fxc that corresponds to the update of the flow within the zone. Similarily, convectiveFluxBC function computes FxcBC, that corresponds to the update of the flow on the borders of the zone: to perform this task, it must take in parameter, in addition to the Primitive value, the value Conservative on every border of the zone. Note that this function is declared as in Sect. 2.

These Fxc and FxcBC updates are combined by the fluxBalance function, that produces the Balance value, which is in turn normalized by the explicitIncrement function into the Rhs value: this value describes how the fluid flow evolves during a step of simulation. Finally, the inlet, outpres and wall functions all produce the Conservative value on a border (which is used by the convectiveFluxBC function), and take in parameter the Conservative value available on the zone related to that border. What distinguish these functions

---

[3] we consider here that missing value constructors are implicitly added in the term signature.

are the guards, which state that they can be used to get the Conservative value only on borders of a specific type.

Generated Rewriting Rules. Figure 5 presents the rewriting rules that can be automatically extracted from the mesh of Fig. 3 (Lines 1–9), and generated from the function specifications in Fig. 4 (Lines 11–30). The rules extracted from the mesh implement the semantics of the zoneOfBorder, getBorderType and bordersOfZone functions. The rules generated from the function specifications state that if a data is currrently not computed by a function (i.e., it is wrapped in the appNone) and is an output of an function, then it is generated by that function[4] (with the corresponding inputs not being generated yet).

```
 1 rl zoneOfBorder(border0) => zone1 .          5 rl getBorderType(border0) => inputFlow .
 2 rl zoneOfBorder(border2) => zone1 .          6 rl getBorderType(border2) => wall .
 3 rl zoneOfBorder(border1) => zone1 .          7 rl getBorderType(border1) => wall .
 4 rl zoneOfBorder(border3) => zone1 .          8 rl getBorderType(border3) => outputFlow .

 9 rl bordersOfZone(zone1) => fieldList(border0, border1, border2, border3) .
10
11 vars Z: zone, B: border .
12
13 rl appNone(data(Primitive, Z)) => appDone(data(Primitive, Z), primitive,
14     appNone(data(Conservative, Z))) .
15 rl appNone(data(Fxc, Z)) => appDone(data(Fxc, Z), convectiveFlux,
16     appNone(data(Primitive, Z))) .
17 rl appNone(data(FxcBC, Z)) => appDone(data(FxcBC, Z), convectiveFluxBC,
18     appNone(data(Primitive, Z)), appNone(data(Conservative, bordersOfZone(Z)))) .
19
20 rl appNone(data(Balance, Z)) => appDone(data(Balance, Z), fluxBalance,
21     appNone(data(Fxc, Z)), appNone(data(FxcB, Z)) ) .
22 rl appNone(data(Rhs, Z)) => appDone(data(Rhs, Z), explicitIncrement,
23     appNone(data(Balance, Z)) ) .
24
25 crl appNone(data(Conservative, B)) => appDone(data(Conservative, B), inlet,
26     appNone(data(Conservative, zoneOfBorder(B)))) if bc_type(B) == Inlet .
27 crl appNone(data(Conservative, B)) => appDone(data(Conservative, B), outpres,
28     appNone(data(Conservative, zoneOfBorder(B)))) if bc_type(B) == Outpres .
29 crl appNone(data(Conservative, B)) => appDone(data(Conservative, B), wall,
30     appNone(data(Conservative, zoneOfBorder(B)))) if bc_type(B) == Wall .
```

Fig. 5. Generated rewriting rules

Rewriting rules Application. The rewriting rules of Figs. 2 and 5 can be used to produce a code assembly computing any data generated by the functions declared in Fig. 4. For instance, to get the code assembly computing the value Rhs on the zone zone1, one needs to rewrite the term appNone(data(Rhs, zone1)). Figure 6 presents the result of this rewriting by only showing the subterms constructed with the data constructors (in green), and the subterms of sort functionID (in blue): we thus obtain a DAG of functions (the subterms of sort functionID) exchanging data. Note that the convectiveFluxBC function has 5 parameters, due to the mesh having 4 borders.

---

[4] We consider here that the name of the function is implicitly declared in the term signature with sort functionID.

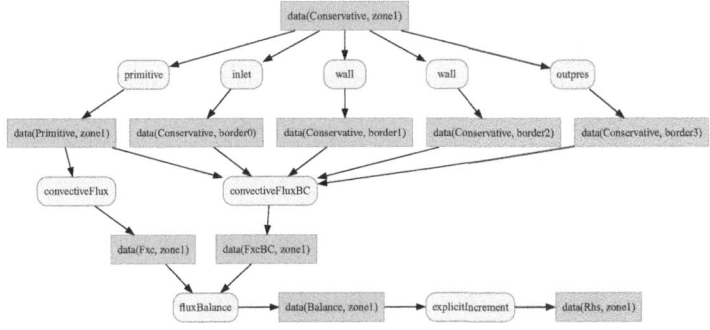

Fig. 6. Generated graph

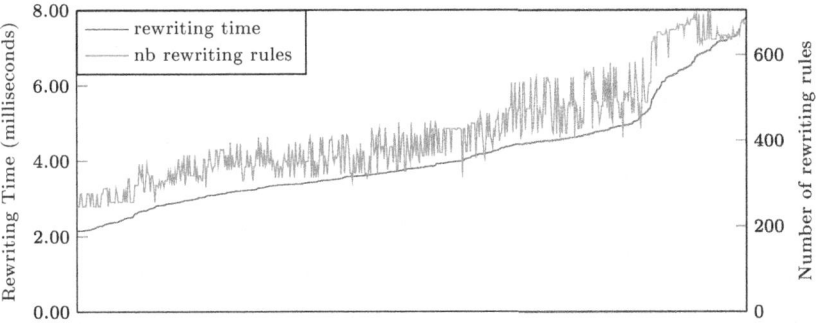

Fig. 7. Rewriting Time and Number of rewriting rules—the x-axis lists the 597 generation problems ordered by rewriting time

5.2 Evaluation of the Library

Each of our 597 generation problems were executed 10 times on a single 2.5 GHz Intel Xeon CPU with 32GB of memory that was hosting a CentOS 8 operating system. Figure 7 presents the execution time and the number of rewriting rules applied for each generation problem : the smallest generation time is 2.14 milliseconds with 246 applied rewriting rules, and the largest generation time is 7.82 milliseconds with 680 applied rewriting rules. On average in this experiment, *hrewrite* applied 104619 rewriting rules per second.

6 Related Work

In the past decades, the subject of ranked and unranked tree transformation has received a lot of attention, especially in the context of XML and internet technologies. XDuce [17] and later CDuce [5] are functional languages with a syntax and type system dedicated to manipulate XML unranked trees. Fast [11] is a tool allowing the composition of tree transducers that implements efficient transformation of ranked trees. Transducers for unranked trees have also been investigated, in particular in the context of efficiency: in [2,14], the authors investigate methods to transform unranked trees in linear time. These approaches are very interesting and could easily be extended with a fixpoint to construct rewriting engines. However such engines would not be expressive enough, in particular they would not be able to correctly manage user-defined rewriting rule conditions. Moreover, as these approaches are focused on trees, they should be adapted to manage DAGs.

As discussed in [15], many existing tools can be used as rewriting engines. In particular, Maude [9] is one of the most efficient. Maude only considers ranked trees, but its very flexible syntax and its support for associativity [13] allow to simulate unranked term constructors whose subterms must all be of the same sort. However, Maude does not support pattern matching of unranked trees, and its terms cannot carry non-term values.

Several term rewriting tools can be accessed directly from a programming language like *hrewrite*. For instance, Maude recently got a new Python API [23] where all declarations can be done using the Maude DSL, and a dedicated API is provided to access specific modules and apply rewriting rules. The `term_rewriting` [24] is a Rust library for rewriting unsorted and ranked terms. The library has a relatively complex API to create terms and rewriting rules directly, but provides a parser to directly extract terms and rewriting rules from their string description. Similarily to Maude in Python, terms of this library cannot directly interact with other Rust data (constructors and variables can only be strings). And the `term-rewriting` [3] is a library for rewriting unsorted and unranked terms implemented in haskell, with a simple API to construct terms and rewriting rules, and where constructors and variable can be any haskell data.

Among the other tools cited in [15], Rascal [6] is a very expressive tool for meta programming that includes pattern matching lists and term rewriting. The application domain of this tool is related to *hrewrite* (which is used to automatically generate code assembly), but it cannot express unranked tree language nor select a specific matching using user-defined guard as it is necessary in our usecase.

7 Conclusion

In this paper, we presented a new rewriting engine on unranked trees, called *hrewrite*. We described the motivations for this new engine, its structure and its Python API.

In future work, we want to extend the functionalities of *hrewrite* by allowing several signatures for a term constructor. Moreover, we intend to improve the efficiency of the rewriting algorithm by investigating tree transducers modulo theory [10]. Indeed, transducers could be an efficient way to implement sets of rewriting rules, and term matching must be performed modulo the subsort relation. Finally, *Computational Fluid Dynamics* is not the only context in which the problem of automatic code assembly occurs, and other contexts may have other expressivity requirement. Consequently, we may investigate the extension of *hrewrite* to other ways to specify data, like row types [22], set types [7] or guarded algebraic data types [26].

Acknowledgements. The author would like to thank Chiara Oberti and the anonymous reviewers for their useful and constructive comments.

References

1. Abate, P., Di Cosmo, R., Treinen, R., Zacchiroli, S.: Dependency solving: A separate concern in component evolution management. J. Syst. Softw. **85**(10), 2228–2240 (2012). https://doi.org/10.1016/J.JSS.2012.02.018
2. Alur, R., D'Antoni, L.: Streaming tree transducers. J. ACM **64**(5), 31:1–31:55 (2017). https://doi.org/10.1145/3092842
3. Avanzini, M., Felgenhauer, B., Sternagel, C., Epifanov, I.: Term-Rewriting. http://cl-informatik.uibk.ac.at/software/haskell-rewriting/, (accessed: 20.03.2024)
4. Becker, G., Scheibel, P., LeGendre, M.P., Gamblin, T.: Managing combinatorial software installations with spack. In: 2016 Third International Workshop on HPC User Support Tools, HUST@SC, pp. 14–23. IEEE Computer Society (2016). https://doi.org/10.1109/HUST.2016.007
5. Benzaken, V., Castagna, G., Frisch, A.: Cduce: an xml-centric general-purpose language. In: Proceedings of the Eighth ACM SIGPLAN International Conference on Functional Programming, ICFP, pp. 51–63. ACM (2003). https://doi.org/10.1145/944705.944711
6. van den Bos, J., Hills, M., Klint, P., van der Storm, T., Vinju, J.J.: Rascal: From algebraic specification to meta-programming. In: Proceedings Second International Workshop on Algebraic Methods in Model-based Software Engineering, AMMSE. EPTCS, vol. 56, pp. 15–32 (2011). https://doi.org/10.4204/EPTCS.56.2
7. Castagna, G.: Programming with union, intersection, and negation types. CoRR **abs/2111.03354** (2021). https://doi.org/10.48550/arXiv.2111.03354
8. Chidlovskii, B.: Using regular tree automata as XML schemas. In: Proceedings of IEEE Advances in Digital Libraries 2000 ADL, pp. 89–98. IEEE Computer Society (2000). https://doi.org/10.1109/ADL.2000.848373
9. Clavel, M., Durán, F., Eker, S., Escobar, S., Lincoln, P., Martí-Oliet, N., Talcott, C.L.: Two decades of maude. In: Logic, Rewriting, and Concurrency—Essays dedicated to José Meseguer on the Occasion of His 65th Birthday. Lecture Notes in Computer Science, vol. 9200, pp. 232–254. Springer, Berlin (2015). https://doi.org/10.1007/978-3-319-23165-5_11
10. D'Antoni, L., Veanes, M.: Automata modulo theories. Commun. ACM **64**(5), 86–95 (2021). https://doi.org/10.1145/3419404
11. D'Antoni, L., Veanes, M., Livshits, B., Molnar, D.: Fast: A transducer-based language for tree manipulation. ACM Trans. Program. Lang. Syst. **38**(1), 1:1–1:32 (2015). https://doi.org/10.1145/2791292
12. Di Cosmo, R., Lienhardt, M., Treinen, R., Zacchiroli, S., Zwolakowski, J., Eiche, A., Agahi, A.: Automated synthesis and deployment of cloud applications. In: ACM/IEEE International Conference on Automated Software Engineering, ASE, pp. 211–222. ACM (2014). https://doi.org/10.1145/2642937.2642980
13. Eker, S.: Associative unification in maude. J. Log. Algebraic Methods Program. **126**, 100747 (2022). https://doi.org/10.1016/J.JLAMP.2021.100747

14. Gallot, P., Lemay, A., Salvati, S.: Linear high-order deterministic tree transducers with regular look-ahead. In: 45th International Symposium on Mathematical Foundations of Computer Science, MFCS. LIPIcs, vol. 170, pp. 38:1–38:13. Schloss Dagstuhl - Leibniz-Zentrum für Informatik (2020). https://doi.org/10.4230/LIPICS.MFCS.2020.38
15. Garavel, H., Tabikh, M., Arrada, I.: Benchmarking Implementations of Term Rewriting and Pattern Matching in Algebraic, Functional, and Object-Oriented Languages: The 4th Rewrite Engines Competition. In: Proceedings of the 12th International Workshop on Rewriting Logic and Its Applications WRLA. LNCS, vol. 11152, pp. 1–25. Springer, Berlin (2018). https://doi.org/10.1007/978-3-319-99840-4_1
16. de Gouw, S., Lienhardt, M., Mauro, J., Nobakht, B., Zavattaro, G.: On the integration of automatic deployment into the ABS modeling language. In: Service Oriented and Cloud Computing—4th European Conference, ESOCC. Lecture Notes in Computer Science, vol. 9306, pp. 49–64. Springer, Berlin (2015). https://doi.org/10.1007/978-3-319-24072-5_4
17. Hosoya, H., Pierce, B.C.: Xduce: A statically typed XML processing language. ACM Trans. Internet Tech. **3**(2), 117–148 (2003). https://doi.org/10.1145/767193.767195
18. Kutsia, T., Marin, M.: Regular expression order-sorted unification and matching. J. Symb. Comput. **67**, 42–67 (2015). https://doi.org/10.1016/J.JSC.2014.08.002
19. Legensky, S., Edwards, D., Bush, R., Poirier, D., Rumsey, C., Cosner, R., Towne, C.: CFD general notation system (CGNS)—Status and future directions, Chap. 752. ARC (2002). https://doi.org/10.2514/6.2002-752
20. Lienhardt, M., Damiani, F., Johnsen, E.B., Mauro, J.: Lazy product discovery in huge configuration spaces. In: ICSE '20: 42nd International Conference on Software Engineering, pp. 1509–1521. ACM (2020). https://doi.org/10.1145/3377811.3380372
21. Lienhardt, M., ter Beek, M.H., Damiani, F.: Product lines of dataflows. J. Syst. Softw. 111928 (2023). https://doi.org/10.1016/j.jss.2023.111928
22. Morris, J.G., McKinna, J.: Abstracting extensible data types: or, rows by any other name. Proc. ACM Program. Lang. **3**(POPL), 12:1–12:28 (2019). https://doi.org/10.1145/3290325
23. Rubio, R.: Maude as a library: An efficient all-purpose programming interface. In: Rewriting Logic and Its Applications—14th International Workshop, WRLA@ETAPS. Lecture Notes in Computer Science, vol. 13252, pp. 274–294. Springer, Berlin (2022). https://doi.org/10.1007/978-3-031-12441-9_14
24. Rule, J.: term_rewriting. https://github.com/joshrule/term-rewriting-rs, (accessed: 20.03.2024)
25. Schwentick, T.: Trees, automata and XML. In: Proceedings of the Twenty-third ACM SIGACT-SIGMOD-SIGART Symposium on Principles of Database Systems, p. 222. ACM (2004). https://doi.org/10.1145/1055558.1055589
26. Simonet, V., Pottier, F.: A constraint-based approach to guarded algebraic data types. ACM Trans. Program. Lang. Syst. **29**(1), 1 (2007). https://doi.org/10.1145/1180475.1180476

A Flexible Framework for Integrating Maude and SMT Solvers Using Python

Geunyeol Yu and Kyungmin Bae(✉)

Pohang University of Science and Technology, Pohang, South Korea
kmbae@postech.ac.kr

Abstract. This paper presents a new implementation of Maude-SE that provides a flexible yet efficient framework for connecting Maude to SMT solvers. There exist previous implementations to integrate Maude and SMT solvers at the C++ level, but they do not support uninterpreted functions and folding reduction, and are very difficult to customize. The new version of Maude-SE supports uninterpreted functions, symbolic reachability analysis with folding, and an abstract Python connector that makes it easy to integrate and customize SMT solving with Maude using its Python API, without having to understand Maude's internal implementation and recompile the source code.

1 Introduction

Rewriting modulo SMT combines term rewriting with SMT solving to provide a powerful framework for modeling and analyzing infinite-state concurrent systems [7,25]. In this approach, system states are symbolically represented as terms constrained by SMT formulas, while transitions between these states are specified using conditional rewrite rules. Rewriting modulo SMT has been widely used in various applications, including security [1,27,28], real-time systems [2–4,21], cyber-physical systems [18–20,24], business process models [15], etc.

The Maude tool [12] provides the basic functionality for SMT solving and symbolic reachability analysis. Maude-SE [30] extends Maude with additional functionality, such as witness generation, for rewriting modulo SMT. However, the existing implementations have several limitations:

- **Limited SMT Theory Support**: only Booleans, integers, and reals are supported, and uninterpreted functions are not supported.
- **Absence of Folding Reduction**: *folding* [5,22], a well-known state-space reduction method for symbolic reachability analysis, is not supported.
- **Lack of Customizability**: altering existing features is very difficult due to the heavy dependence on Maude's internal C++ implementation.

To address these difficulties, this paper presents a new implementation of Maude-SE that provides a flexible yet efficient framework for connecting Maude to SMT solvers. The new version of Maude-SE supports uninterpreted functions and symbolic reachability analysis with folding. A key feature of our tool is an abstract Python connector that facilitates integrating Maude to SMT solvers and customizing its functionality at the Python level, without having to understand Maude's internal implementation and recompile the source code.

Developing abstract Python connectors that achieve both high flexibility and high performance is a difficult design challenge. We use the `maude` library [26] to invoke user-defined Python functions directly from within Maude. However, using the `maude` library requires an understanding of the internal data structures of Maude, making this a difficult task. To achieve the best performance of symbol reachability analysis with folding, term manipulation and symbol search graph construction must be implemented at the C++ level.

Our tool has been implemented to enjoy the best of both worlds: flexible capability of SMT solving using each SMT solver's Python API, and efficient exploration of symbolic search space using Maude. Our abstract connector hides details about Maude's internal implementation, so users only need to understand the Python API for the target SMT solver. The main routine for symbolic reachability analysis and folding are implemented at the C++ level and use the `maude` library to invoke user-defined Python functions for SMT solving.

In addition to folding and abstract connectors, the new version of Maude-SE supports improved modeling and analysis capabilities as follows:

– We have integrated four widely used SMT solvers with Maude: CVC4 [9], CVC5 [8], Yices2 [16], and Z3 [13]. Note that Maude only supports CVC4 and Yices2, and the older version of Maude-SE additionally supports Z3.
– We support the theory of equality and uninterpreted functions, which is not supported in the previous implementations. This involves a Maude-level interface to declare any free symbols as SMT symbols.

The rest of the paper is organized as follows. Section 2 gives some background on rewriting modulo SMT. Section 3 introduces the usage of Maude-SE. Section 4 presents the design and implementation of Maude-SE, including a case study of connecting a new SMT solver to Maude using an abstract Python connector. Finally, Sect. 5 gives some concluding remarks. The tool, together with several examples, is available at https://maude-se.github.io.

2 Background on Rewriting Modulo SMT

An *order-sorted signature* is a triple $\Sigma = (S, \leq, F)$ with a finite poset of sorts $(S, \leq)$ and a set of function symbols F typed with sorts in $(S, \leq)$. We denote by $\mathcal{T}_\Sigma(X)_s$ the set of Σ-terms of sort s over a set X of S-sorted variables, and by $\mathcal{T}_{\Sigma,s}$ the set of ground Σ-terms of sort s. $\mathcal{T}_\Sigma(X)$ and $\mathcal{T}_\Sigma$ denote the set of Σ-terms and the set of ground Σ-terms, respectively. The set of variables in a term t is denoted by $var(t)$.

A substitution $\theta : X \to \mathcal{T}_\Sigma(X)$ maps each variable to a term of the same sort, where $\theta(x) \neq x$ for only finitely many variables $x \in X$. The *domain* of a substitution θ is a finite set $dom(\theta) = \{x \in X \mid \theta(x) \neq x\}$.

A *built-in subtheory* $\mathcal{E}_0$ is a first-order theory that is handled by SMT solving. The built-in subsignature $\Sigma_0 = (S_0, F_0)$ of $\mathcal{E}_0$ is a subsignature of Σ, where any Σ-terms of built-in sorts are built-in terms (i.e., $\mathcal{T}_\Sigma(X_0)_{s_0} = \mathcal{T}_{\Sigma_0}(X_0)_{s_0}$ for each $s_0 \in \Sigma_0$). The set $QF_{\Sigma_0}(X_0)$ denotes the set of quantifier-free Σ_0-formulas over $X_0 \subseteq X$ a set of built-in variables.

A *rewrite theory modulo a built-in subtheory* $\mathcal{E}_0$ [25] is a triple $\mathcal{R} = (\Sigma, E, R)$, where: (i) Σ is an order-sorted signature with the built-in subsignature $\Sigma_0 \subseteq \Sigma$; (ii) (Σ, E) is an equational theory with E a set of Σ-equations; and (iii) R is a set of topmost[1] rewrite rules of the form $l \longrightarrow r$ **if** ϕ, with $l, r \in \mathcal{T}_\Sigma(X)_{State}$ for some sort *State*, and $\phi \in QF_{\Sigma_0}(X_0)$. We require that (Σ, E) *protects* $\mathcal{E}_0$(i.e., for built-in terms $t_1, t_2 \in \mathcal{T}_{\Sigma_0}$, $t_1 =_E t_2$ implies $\mathcal{E}_0 \models t_1 = t_2$).

A *constrained term* is a pair $(t; \phi)$ of a term $t \in \mathcal{T}_\Sigma(X_0)_{State}$ and a constraint $\phi \in QF_{\Sigma_0}(X_0)$. Intuitively, $(t; \phi)$ represents the set of all ground instances $t\rho$ of term t by substitution ρ such that $\mathcal{E}_0 \models \phi\rho$ holds. A *one-step symbolic rewrite* $(t; \phi_t) \rightsquigarrow_\mathcal{R} (u; \phi_u)$ holds iff there exists a rule $l \longrightarrow r$ **if** ϕ and a substitution θ such that (i) $l\theta =_E t$, (ii) $r\theta =_E u$, (iii) $\mathcal{E}_0 \models \phi_u$, where $\mathcal{E}_0 \models (\phi_t \wedge \phi\theta) \leftrightarrow \phi_u$ [6,7,25].

The correctness of rewriting modulo SMT [25] guarantees that symbolic rewrites have corresponding concrete rewrites, and vice versa.

An *abstraction of built-ins* for a term $t \in \mathcal{T}_\Sigma(X)$ is a pair (t°, σ°) of a term $t^\circ \in \mathcal{T}_{\Sigma \setminus \Sigma_0}(X)$ and a substitution $\sigma^\circ : X_0 \to \mathcal{T}_{\Sigma_0}(X_0)$ such that $t = t^\circ \sigma^\circ$ and t° contains no duplicate built-in variables. Any non-variable built-in subterms of t are replaced by distinct built-in variables in t°. Let $(t; \phi)$ be a constrained term and (t°, σ°) an abstraction of built-ins for t. If $dom(\sigma^\circ) \cap var((t; \phi)) = \emptyset$, then $(t; \phi)$ and $(t^\circ; \phi \wedge \Psi_{\sigma^\circ})$ are equivalent, where $\Psi_{\sigma^\circ} = \bigwedge_{x \in dom(\sigma^\circ)} x = x\sigma^\circ$ [25].

Given two constrained terms $(t; \phi_t)$ and $(u; \phi_u)$, a *subsumption relation* $(t; \phi_t) \sqsubseteq (u; \phi_u)$ holds iff there exists a substitution θ such that (i) $t =_E u\theta$, and (ii) $\mathcal{E}_0 \models \phi_t \to \phi_u \theta$ [22]. If $(t; \phi_t) \sqsubseteq (u; \phi_u)$, then the set of terms represented by $(u; \phi_u)$ includes the set of terms represented by $(t; \phi_t)$.

Maude [12] provides an interface to perform symbolic reachability analysis using connections to Yices2 [16] and CVC4 [9]. It declares built-in signatures with three sorts `Boolean`, `Integer`, and `Real` for the theories of Booleans, integers, and reals, respectively, in the SMT-LIB standard [10]. The `check` command invokes the underlying SMT solver to check the satisfiability of an SMT formula. The `smt-search` command performs symbolic reachability analysis.

3 The Usage of Maude-SE

This section describes the user interface, commands and usage of Maude-SE. Section 3.1 briefly explains the admissibility requirements that system modules should meet in Maude-SE. Section 3.2 explains the SMT theories supported by

---

[1] To ensure that all rewrite rules take place at the top of the term, we assume that sort *State* is at the top of one of the connected component of poset of sorts, and no operator in Σ has *State* or any of its subsorts as an argument sort.

Maude-SE, including the theory of equality and uninterpreted functions (EUF). Section 3.3 explains Maude-SE's analysis commands.

3.1 Admissibility Requirements

In Maude-SE, a system module M specifies a rewrite theory modulo a built-in subtheory satisfying the following conditions:

- The equational theory of M protects the underlying SMT theory; and
- For each rule $l \longrightarrow r$ **if** ϕ, any variable that does not appear in l has a built-in sort, and ϕ is decomposed into pure SMT formulas and non-SMT conditions, where non-SMT conditions are admissible in the usual way [12].

As an example, we consider the following specification adapted from [2]. The system module COFFEE-MACHINE specifies the coffee machine parametric timed automaton (PTA). The behavior of COFFEE-MACHINE is the same as the original one [2], except that we use auxiliary functions and equations, where each rule condition is decomposed into an non-SMT condition and a pure SMT formula.

```
mod COFFEE-MACHINE is
  protecting REAL .
  sorts State Location .
  op <_;_;_> <_;_;_> : Location Real Real Real Real Real -> State [ctor] .
  op [_;_;_] <_;_;_> : Location Real Real Real Real -> State [ctor] .
  ops idle addsugar preparingcoffee cdone : -> Location [ctor] .

  sort Tuple{Location,Real,Real,Boolean} .
  op {_,_,_,_} : Location Real Real Boolean -> Tuple{Location,Real,Real,Boolean} [ctor] .

  vars T X Y X' Y' P1 P2 P3 : Real .    vars L L' : Location .    var PHI : Boolean .

  crl [tick] : [ L ; X ; Y ] < P1 ; P2 ; P3 > => < L ; X + T ; Y + T > < P1 ; P2 ; P3 >
    if PHI := tickCond(L, T, X, Y, P2, P3) /\ (T >= 0/1 and PHI) = true [nonexec] .
  crl [toAddsugar] : < L ; X ; Y > < P1 ; P2 ; P3 > => [ L' ; X' ; Y' ] < P1 ; P2 ; P3 >
    if {L', X', Y', PHI} := nextAddSugar(L, X, Y, P1, P2) /\ PHI = true .
  crl [toOther] : < L ; X ; Y > < P1 ; P2 ; P3 > => [ L' ; X' ; Y' ] < P1 ; P2 ; P3 >
    if {L', X', Y', PHI} := nextOther(L, X, Y, P2, P3) /\ PHI = true .

  op tickCond : Location Real Real Real Real Real -> Boolean .
  eq tickCond(idle, T, X, Y, P2, P3) = true .
  eq tickCond(addsugar, T, X, Y, P2, P3) = Y + T <= P2 .
  eq tickCond(preparingcoffee, T, X, Y, P2, P3) = Y + T <= P3 .
  eq tickCond(cdone, T, X, Y, P2, P3) = X + T <= 10/1 .

  op nextAddSugar : Location Real Real Real Real -> Tuple{Location,Real,Real,Boolean} .
  eq nextAddSugar(idle, X, Y, P1, P2) = {addsugar, 0/1, 0/1, 0/1 <= P2} .
  eq nextAddSugar(addsugar, X, Y, P1, P2) = {addsugar, 0/1, Y, Y <= P2 and X >= P1} .
  eq nextAddSugar(cdone, X, Y, P1, P2) = {addsugar, 0/1, 0/1, 0/1 <= P2} .

  op nextOther : Location Real Real Real Real -> Tuple{Location,Real,Real,Boolean} .
  eq nextOther(addsugar, X, Y, P2, P3) = {preparingcoffee, X, Y, Y <= P3 and Y === P2} .
  eq nextOther(preparingcoffee, X, Y, P2, P3) = {cdone, 0/1, Y, 0/1 <= 10/1 and Y === P3} .
  eq nextOther(cdone, X, Y, P2, P3) = {idle, X, Y, X === 10/1} .
endm
```

3.2 Supported SMT Theories

The new version of Maude-SE supports the theories of Booleans, integers, and reals, which are already supported in the previous implementations, declared in the functional modules BOOLEAN, INTEGER, and REAL, respectively. In addition, we support the theory of equality and uninterpreted functions (EUF), which is widely used for typical SMT-based modeling and analysis.

Following the existing method of declaring SMT symbols in Maude [12], SMT symbols in Maude-SE are declared using the operator attribute special (id-hook SMT_Symbol (*SMTSymbolId*)), where *SMTSymbolId* represents the corresponding symbol in the underlying SMT theory.

For example, in the following module, we declare an uninterpreted function f with sort A, an uninterpreted binary function g with integer arguments, and an equality symbol for sort A, where EUF denotes uninterpreted symbols, and === denotes the equality symbol in the SMT theory.

```
fmod EUF-EX is
    protecting INTEGER .
    sort A .
    op f : A -> A [special (id-hook SMT_Symbol (EUF))] .
    op g : Integer Integer -> Integer [special (id-hook SMT_Symbol (EUF))] .
    op _===_ : A A -> Boolean [special (id-hook SMT_Symbol (===))] .
endfm
```

There is some admissibility requirement for uninterpreted functions, which is assumed to protect the corresponding SMT theory. Any operator declared as an uninterpreted SMT function must be a free constructor with respect to the equational theory. That is, there should be no equational axioms (e.g., assoc, comm, and id) on operators for uninterpreted SMT functions, and there is no equation to reduce such uninterpreted functions.

3.3 Analysis Commands

Maude-SE provides various analysis commands, including check, show model, smt-search, and show smt-path, using Maude's meta-interpreter [12].[2] The check command determines the satisfiability of Boolean formulas. The show model command returns the satisfying assignment, if any, for the last check command. The smt-search command performs symbolic reachability analysis, potentially with folding. Finally, the show smt-path command displays a path for the last smt-search result.

Given a module M, a Boolean formula ψ, and an (optional) SMT theory Th, the check command determines the satisfiability of ψ under the theory Th:[3]

$$\text{check in } M \ : \ \psi \text{ using } Th \ .$$

---

[2] As usual, each command has a corresponding metalevel function implemented at the C++ level, and the meta-interpreter invokes these metalevel functions.

[3] For SMT solvers with automatic theory detection, 'using Th' can be omitted.

As an example, consider the EUF-EX module of Sect. 3.2. We can check the satisfiability of formulas and get a satisfying assignment as follows.

```
MaudeSE> check I:Integer > 2 and g(I:Integer, J:Integer) > 3 and f(X:A) === X:A .
result: sat

MaudeSE> show model .
assignment: I |--> 3; J |--> 2; X |--> A!val!0; g |--> [else -> 4]; f |--> [else -> A!val!0]
```

The smt-search command performs symbolic reachability analysis, along with folding. Given an initial term $t \in T_\Sigma(X_0)$, a goal pattern $u \in T_\Sigma(X)$, and a goal condition $\varphi \in QF_{\Sigma_0}(X_0)$, the following command searches for n solutions that are reachable within m rewrite steps from t, match the goal pattern u, and satisfy the goal condition φ under the SMT theory Th:

$$\{\text{fold}\}\ \text{smt-search}\ [n, m]\ \text{in}\ M\ :\ t\ =>*\ u\ \text{such that}\ \varphi\ \text{using}\ Th\ .$$

where {fold}, Th, n, and m are optional. With {fold}, the command uses *folding* [5,22] to reduce the state space by ignoring subsumed states in terms of the subsumption relation $\sqsubseteq$ between constrained terms (see Sect. 2).

For example, consider the system module COFFEE-MACHINE in Sect. 3.1. The following command finds the first solution of the coffee machine that goes to cdone with all the clocks and parameters are greater than or equal to 0. (A symbolic term and an SMT constraint represent a constrained term for a goal. A concrete term represents a concrete witness of the constrained term, given by a satisfying assignment obtained by the underlying SMT solver.) It is worth noting that the same command without folding does not terminate.

```
MaudeSE> {fold} smt-search [1]: < idle    ; X:Real  ; Y:Real   > < P1:Real ; P2:Real ; P3:Real >
                            =>* < cdone ; X':Real ; Y':Real > < P1:Real ; P2:Real ; P3:Real >
               such that X:Real >= 0/1 and Y:Real >= 0/1 and X':Real >= 0/1 and Y':Real > 0/1
                      and P1:Real >= 0/1 and P2:Real >= 0/1 and P3:Real >= 0/1 and P1 >= P2 .
Solution 1 (state 6)
Symbolic term: < cdone ;
Constraint: X >= 0/1 and Y >= 0/1 ...
Substitution:
Concrete term: < cdone ; 0/1 ; 0/1 > < 1/2 ; 0/1 ; 0/1 >
```

The show smt-path command returns a path for the last smt-search result. It takes two arguments: a path type (either symbolic or concrete) and a state number. A symbolic path is given by a sequence of constrained terms and rewrite rules. The corresponding concrete path is an instance of the symbolic path with a satisfying assignment. For example, the following commands show the symbolic and concrete paths for the above smt-search command.

```
MaudeSE> show smt-path symbolic 6 .
( X >= 0/1 and Y >= 0/1 ... ; < idle ; X ; Y > < P1 ; P2 ; P3 > ) =====[ toAddsugar ]=====>
( X >= 0/1 and Y >= 0/1 ... ; [addsugar ; 0/1 ; 0/1] < P1 ; P2 ; P3 > ) ...

MaudeSE> show smt-path concrete 6 .
< idle ; 0/1 ; 0/1 > < 1/2 ; 0/1 ; 0/1 > ===[ toAddsugar ]===>
[addsugar ; 0/1 ; 0/1] < 1/2 ; 0/1 ; 0/1 > ...
```

4 Design and Implementation of Maude-SE

This section presents the design and implementation of Maude-SE. We extend the existing Maude-SMT wrapper interface with additional functionality such as folding, uninterpreted functions, and concrete witness generation. We also propose an abstract Python connector to easily implement and customize new connectors without having to understand Maude's internal implementation.

4.1 Overall Architecture

Maude-SE Wrapper Interface.
The original Maude implementation defines a generic SMT wrapper interface for connecting SMT solvers to Maude. It provides several functions, such as checkSat for checking satisfiability using the SMT solver, maude2smt for converting Maude terms to data structures for the SMT solver, etc. However, it has several limitations as mentioned in Sect. 1.

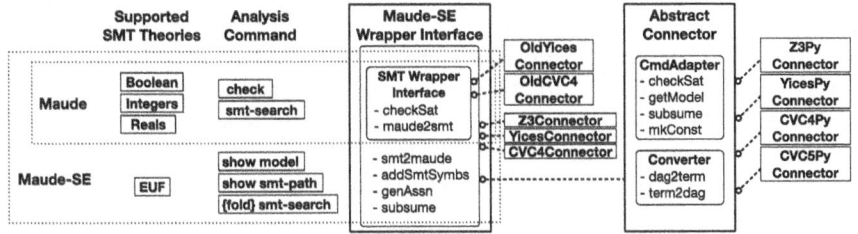

Fig. 1. The architecture of Maude-SE

We extend the Maude SMT wrapper interface to provide additional features and functionalities, including uninterpreted functions and symbolic reachability analysis with folding. Figure 1 illustrate an overall architecture of the Maude-SE wrapper interface. This interface allows various SMT solvers, such as Z3, Yices2 and CVC4, to be connected to Maude.

The Maude-SE wrapper interface defines four additional functions as follows: smt2maude converts the data structures for the SMT solver into Maude terms; genAssn returns a satisfying assignment; subsume computes the subsumption relation; and addSmtSymbs associates SMT operators in Maude, declared with the attribute SMT_Symbol, with the corresponding SMT symbols.

The user interface and commands, explained in Sect. 3, are implemented using this new wrapper interface. E.g., show model and show smt-path are implemented using concrete witness generation with smt2maude and genAssn, and EUF is supported with addSmtSymbs. The smt-search command has been reimplemented to support folding with subsume.

Abstract Python Connector. Adding a new SMT connector or customizing such connectors is difficult, because it requires an understanding of both Maude's internal implementation and the SMT solver APIs. For example, implementing `checkSat` requires manipulating the Maude term data structure. Furthermore, the Maude source files must be compiled each time an update is made

To facilitate adding and customizing connectors, we introduce an abstract connector at the Python level. Our abstract connector is implemented in C++, but invokes user-defined Python functions for SMT solving directly from within Maude. Thus, an instance of the abstract connector can be easily customized by modifying its Python code without recompiling the source files.

As shown in Fig. 1, our abstract connector has two Python components. `CmdAdapter` provides functions for SMT solving that are independent of Maude and define what the SMT solver should do. `Converter` provides functions to translate Maude terms into data structures for the SMT solver and vice versa. This part is intended to be implemented using the `maude` library [26].

The component `CmdAdapter` provides four functions. The function `checkSat` checks the satisfiability of an SMT formula, and `getModel` returns a satisfying assignment if one exists; these functions provide basic SMT-solving capabilities. For two constrained terms $(t_1\,;\phi_1)$ and $(t_2\,;\phi_2)$, the function `subsume` checks if $\phi_1 \to \phi_2\theta$ is valid, where θ denotes a matching substitution such that $t_1 =_E t_2\theta$ (which is obtained by the abstract connector). The function `mkConst` builds an SMT formula to be stored in the state space, given an accumulated SMT constraint and a newly generated constraint.

The component `Converter` provides two translation functions. The `dag2term` function translates Maude terms into data structures for the SMT solver, and `term2dag` does the opposite. The Maude-dependent parts of these functions can be implemented in Python using the `maude` library.

Using the abstract Python connector, we have implemented connectors for widely used SMT solvers, including Z3, Yices, CVC4, and CVC5.

4.2 Case Study: Connecting a Z3 Solver

This section demonstrates the flexibility of our interface with a case study on connecting a Z3 solver. We first show a simple implementation for the two Python components `CmdAdapter` and `Converter`, and then explain how to customize the implementation of `CmdAdapter` for different purposes.

Implementing CmdAdapter. The following class `Z3CmdAdapter` implements the base abstract class `CmdAdapter`. A Z3 solver object, assigned to `self.solver`, is used to check the satisfiability of the input formulas (`checkSat`) and to build a satisfying assignment (`getModel`). We store an SMT formula in the state space as is (`mkConst`). For two constrained terms $(t\,;\phi_t)$ and $(u\,;\phi_u)$ and a substitution θ such that $t =_E u\theta$, `subsume`(θ, ϕ_t, ϕ_u) checks the validity of $\phi_t \to \phi_u\theta$.

```
class Z3CmdAdapter(CmdAdapter):
    def checkSat(self, *consts):
        self.solver = z3.Solver()
```

```
        self.solver.add(*consts)
        return self.solver.check() == z3.sat

    def getModel(self):
        return {v : self.solver.model()[v] for v in self.solver.model().decls()}

    def mkConst(self, cur, acc):
        return z3.And(cur, acc)

    def subsume(self, subst, cur, prev):
        tmpSolver = z3.Solver()
        tmpPhi = z3.substitute(prev, *[(p, subst[p]) for p in subst])
        tmpSolver.add(z3.Not(z3.Implies(cur, tmpPhi)))
        return tmpSolver.check() == z3.unsat
```

Implementing `Converter`. The following class `Z3Converter` implements the base abstract class `Converter`. In the constructor, we initialize four Python maps: `_symb_map` and `_const_map` are used to translate Maude symbols, and `_sort_map` and `_u_sort` store Maude's sort information for translation.

```
class Z3Converter(Converter):
    def __init__(self):
        self._symb_map  = { "not_" : z3.Not, ..., "_<_"   : z3.z3.ArithRef.__lt__, ... }
        self._const_map = { "true" : z3.BoolVal, "false" : z3.BoolVal,
                            "<Integers>" : z3.IntVal, "<Reals>" : z3.RealVal, ... }
        self._sort_map  = { "Integer" : z3.IntSort, "Real" : z3.RealSort,
                            "Boolean" : z3.BoolSort, ... }
        self._u_sort = dict()     # a map for uninterpreted sort
```

The `dag2term` function recursively builds a Z3 data structure from a Maude term representing an SMT formula, using two auxiliary functions `_declSort` and `_declFunc`. The `dag2term` function takes an additional argument `special`, which contains the theory name and sorts of each uninterpreted function symbol.

```
def dag2term(self, t: Term, special: Dict):
    if t.isVariable():
        v_sort, v_name = str(t.getSort()), t.getVarName()
        return z3.Const(v_name, self._declSort(v_sort))
    symbol, symbol_sort = str(t.symbol()), str(t.getSort())
    theory, sorts = special[symbol]

    if theory == "euf" :
        args = [self.dag2term(arg, special) for arg in t.arguments()]
        f = self._declFunc(symbol, *sorts)     # declare the function
        return f(*args)
    if symbol in self._symb_map:
        op = self._symb_map[symbol]
        args = [self.dag2term(arg, special) for arg in t.arguments()]
        return op(*args)
    if symbol in self._const_map:
        val = str(t)
        ...  # cleanup the "val" string
        return self._const_map[symbol](val)

def _declSort(self, sort: str):
    return self._u_sort[sort]() if sort in self._u_sort else z3.DeclareSort(sort)

def _declFunc(self, func: str, *args):
    return z3.Function(func, *args)
```

The `term2dag` function converts a Z3 data structure representing an SMT formula into a Maude term. We use the `parseTerm` function from the `maude` library, which parses a string and builds a Maude term. For this purpose, the auxiliary function _toString creates a string from a Z3 data structure.

```
def term2dag(self, term, module):
    return module.parseTerm(self._toString(term))

def _toString(self, term):
    if z3.is_and(term):
        return " and ".join([self._toString(t) for t in term.children()])
    ...
    if z3.is_add(term):
        l, r = self._toString(term.arg(0)), self._toString(term.arg(1))
        return f"{l} + {r}"
    if isinstance(term, z3.z3.FuncDeclRef): # variable
        sort_table = {"Int" : "Integer", "Real" : "Real", "Bool" : "Boolean"}
        sort_s = str(term.range())
        return f"{term.name()}:{sort_table[sort_s]}"
    if isinstance(term, z3.RatNumRef):     # rational
        return f"({term.numerator()}/{term.denominator()}).Real"
    ...
    if isinstance(term, z3.z3.ArithRef): # In this case, the term must be a variable
        return f"{term}:Integer" if term.is_int() else f"{term}:Real"
```

Customizing `CmdAdapter`. The following shows three variations of the `mkConst` function implemented using different Z3 tactics. The first function uses rewriting to obtain a simplified formula. The other functions use a syntactic/semantic equality check to remove subformulas that are subsumed by context

```
def mkConst(self, cur, acc):  # version 1: applying rewriting simplification
    return z3.simplify(z3.And(cur, acc))

def mkConst(self, cur, acc):  # version 2: removing sub-formulas, subsumed by context
    return z3.Tactic('ctx-simplify').apply(z3.And(cur, acc)).as_expr()

def mkConst(self, cur, acc):  # version 3: using solver to check context subsumption
    return z3.Tactic('ctx-solver-simplify').apply(z3.And(cur, acc)).as_expr()
```

The following shows two variants of the `checkSat` function implemented using the Z3 tactics described in [23]. The first function simplifies an input formula and checks if the formula reduces to `true`. The other function first applies Gaussian elimination before checking the satisfiability of a formula.

```
def checkSat(self, *consts):  # version 1: checkSat with simplification only
    self.solver = z3.Tactic('simplify').solver()
    self.solver.add(*consts)
    return self.solver.check() == z3.sat

def checkSat(self, *consts):  # version 2: checkSat with Gaussian elimination and SMT
    self.solver = z3.Then('solve-eqs', 'smt').solver()
    self.solver.add(*consts)
    return self.solver.check() == z3.sat
```

4.3 Implementation of `smt-search`

This section explains the main idea and data structures used to implement the `smt-search` command with folding. We also explain a theory transformation that allows system modules with the more relaxed admissibility requirements shown in Section 3.1 than the original admissibility requirements in [25].

Theory Transformation. As explained in [25], to match a term to a pattern u modulo SMT, we need to obtain an abstraction of built-ins (u°, σ°) for u.[4] There are two cases where matching modulo SMT is performed: applying rewrite rules and checking subsumption for folding. We maintain each constrained term in the state space as its abstraction of builtins to avoid repeated computations.

For this purpose, we consider a theory transformation, introduced in [25]. Given a rewrite theory $\mathcal{R}$, we obtain an equivalent rewrite theory $\mathcal{R}^\circ$ without duplicate built-in variables on the left side of each rule. More precisely, each rule $l \longrightarrow r$ **if** ϕ in $\mathcal{R}$ is transformed into the following rule in $\mathcal{R}^\circ$, where (l°, σ°) is an abstraction of built-ins for l and $\Psi_{\sigma^\circ} = \bigwedge_{x \in dom(\sigma^\circ)} x = x\sigma^\circ$:

$$l^\circ \longrightarrow r \text{ if } \phi \wedge \Psi_{\sigma^\circ}$$

Consider a constrained term $(t^\circ\,;\phi_t)$ and a topmost rule $l^\circ \longrightarrow r$ **if** $\psi \wedge \Psi$, where t° and l° do not contain duplicate built-in variables, ψ denotes a non-SMT condition, and Ψ denotes a pure SMT formula. Suppose the rule is applied to t° and a term $u = r\theta$ is obtained with substitution θ such that $t^\circ =_E l^\circ \theta$. We first obtain a term v by renaming each built-in variable in u that appears in r and Ψ but not in l°. We then store the following constrained term into the state space, where (v°, σ°) is an abstraction of built-ins for v:

$$(v^\circ\,;\phi_t \wedge \Psi\theta \wedge \Psi_{\sigma^\circ})$$

Folding. Suppose there are N constrained terms in the previously explored state space. Consider a new constrained term $(t;\phi)$. A naive implementation of folding compares $(t\,;\phi)$ with all N constrained terms, resulting in a time complexity of $O(N^2)$ for state space exploration, which is $O(N)$ without folding. Also, some of these N constrained terms may have "equivalent" patterns up to renaming; if t does not match one of these patterns, then t does not match all of them.

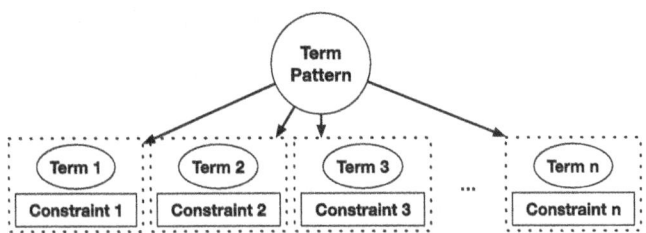

Fig. 2. The data structure for efficient folding

We implement a data structure to concisely maintain constrained terms with equivalent patterns. As shown in Fig. 2, it is a single-depth tree where the root is a representative pattern u and the children are constrained terms with a pattern

---

[4] To illustrate, consider two terms $f(1)$ and $f(x+1)$, where 1 and $x + 1$ are terms of built-in sorts. $f(1)$ does not match $f(x+1)$, but for $\theta = \{x \mapsto 0\}$, $f(1)$ and $f(x+1)\theta$ are equivalent modulo SMT. Using the abstractions of built-ins $(f(y), \{y \mapsto 1\})$ and $(f(z), \{z \mapsto x+1\})$, we can easily check $(f(y)\,;y = 1) \sqsubseteq (f(z)\,;z = x+1)$.

equivalent to u, We use this structure to avoid redundant matching: for a new constrained term $(t\,;\phi)$, if t does not match the representative pattern, then we can ignore all constrained terms in the tree.

We also use Maude's "folder" data structure for narrowing-based reachability analysis [14]. Given a set of terms, it identifies and manages a set of the most general patterns in terms of E-matching. For example, if a term t matches a term u, only u is stored in the data structure. This data structure is part of the official Maude distribution, and is well maintained and efficiently implemented. We utilize it to minimize the number of term matchings required for folding.

5 Conclusion

We have presented an improved implementation of Maude-SE that provides a flexible yet efficient framework for connecting Maude to SMT solvers. In the previous implementations, uninterpreted functions are not supported, symbolic reachability analysis is not integrated with folding, and customizing SMT solving features requires direct modification of Maude's internal C++ implementation. The new version of Maude-SE now supports uninterpreted functions, symbolic reachability analysis with folding, and abstract Python connectors to easily add new SMT solvers and customize its functionality in Python. Our implementation has successfully addressed the design challenges of achieving both high flexibility and high performance, combining flexible capability of SMT solving in Python and efficient exploration of symbolic search space in Maude.

Future work includes connecting Maude to various SMT solvers, such as dReal [17] and MathSat [11], using our Python connectors, supporting arbitrary SMT theories with a user-level interface, and implementing search strategies for incremental rewriting modulo SMT [29].

Acknowledgments. This work was supported in part by the National Research Foundation of Korea (NRF) grants funded by the Korea government (MSIT) (No. 2021R1A5A1021944 and No. RS-2023-00251577), and by the NATO Science for Peace and Security Programme project SymSafe (grant number G6133).

References

1. Aires Urquiza, A., Alturki, M.A., Ban Kirigin, T., Kanovich, M., Nigam, V., Scedrov, A., Talcott, C.: Resource and timing aspects of security protocols. J. Comput. Secur. **29**(3), 299–340 (2021). https://doi.org/10.3233/JCS-200012
2. Arias, J., Bae, K., Olarte, C., Ölveczky, P.C., Petrucci, L., Rømming, F.: Rewriting logic semantics and symbolic analysis for parametric timed automata. In: International Workshop on Formal Techniques for Safety-Critical Systems, pp. 3–15. ACM (2022). https://doi.org/10.1145/3563822.3569923
3. Arias, J., Bae, K., Olarte, C., Ölveczky, P.C., Petrucci, L., Rømming, F.: Symbolic analysis and parameter synthesis for time Petri nets using Maude and SMT solving. In: International Conference on Applications and Theory of Petri Nets and Concurrency. LNCS, vol. 13929, pp. 369–392. Springer, Cham (2023). https://doi.org/10.1007/978-3-031-33620-1_20

4. Arias, J., Bae, K., Olarte, C., Ölveczky, P.C., Petrucci, L., Rømming, F.: Symbolic analysis and parameter synthesis for networks of parametric timed automata with global variables using Maude and SMT solving. Sci. Comput. Programm. **233**, 103074 (2024). https://doi.org/10.1016/j.scico.2023.103074
5. Bae, K., Escobar, S., Meseguer, J.: Abstract logical model checking of infinite-state systems using narrowing. In: International Conference on Rewriting Techniques and Applications. LIPIcs, vol. 21, pp. 81–96. Schloss Dagstuhl – Leibniz-Zentrum für Informatik (2013). https://doi.org/10.4230/LIPIcs.RTA.2013.81
6. Bae, K., Rocha, C.: Guarded terms for rewriting modulo SMT. In: International Conference on Formal Aspects of Component Software. LNCS, vol. 10487, pp. 78–97. Springer, Cham (2017). https://doi.org/10.1007/978-3-319-68034-7_5
7. Bae, K., Rocha, C.: Symbolic state space reduction with guarded terms for rewriting modulo SMT. Sci. Comput. Program. **178**, 20–42 (2019). https://doi.org/10.1016/j.scico.2019.03.006
8. Barbosa, H., Barrett, C., Brain, M., Kremer, G., Lachnitt, H., Mann, M., Mohamed, A., Mohamed, M., Niemetz, A., Nötzli, A., et al.: cvc5: A versatile and industrial-strength SMT solver. In: International Conference on Tools and Algorithms for the Construction and Analysis of Systems. LNCS, vol. 13243, pp. 415–442. Springer, Cham (2022). https://doi.org/10.1007/978-3-030-99524-9_24
9. Barrett, C., Conway, C.L., Deters, M., Hadarean, L., Jovanović, D., King, T., Reynolds, A., Tinelli, C.: CVC4. In: International Conference on Computer Aided Verification. LNCS, vol. 6806, pp. 171–177. Springer, Berlin (2011). https://doi.org/10.1007/978-3-642-22110-1_14
10. Barrett, C., Stump, A., Tinelli, C.: The SMT-LIB standard: Version 2.0. Technical report, Department of Computer Science, The University of Iowa (2010). Available at www.SMT-LIB.org
11. Cimatti, A., Griggio, A., Schaafsma, B.J., Sebastiani, R.: The MathSAT5 SMT solver. In: International Conference on Tools and Algorithms for the Construction and Analysis of Systems. LNCS, vol. 7795, pp. 93–107. Springer, Berlin (2013). https://doi.org/10.1007/978-3-642-36742-7_7
12. Clavel, M., Durán, F., Eker, S., Escobar, S., Lincoln, P., Martí-Oliet, N., Meseguer, J., Rubio, R., Talcott, C.: Maude manual (version 3.3.1). Technical report SRI International, Menlo Park (2023)
13. De Moura, L., Bjørner, N.: Z3: An efficient SMT solver. In: International Conference on Tools and Algorithms for the Construction and Analysis of Systems. LNCS, vol. 4963, pp. 337–340. Springer, Berlin (2008). https://doi.org/10.1007/978-3-540-78800-3_24
14. Durán, F., Eker, S., Escobar, S., Martí-Oliet, N., Meseguer, J., Rubio, R., Talcott, C.: Equational unification and matching, and symbolic reachability analysis in Maude 3.2 (system description). In: International Joint Conference on Automated Reasoning. LNCS, vol. 13385, pp. 529–540. Springer, Cham (2022). https://doi.org/10.1007/978-3-031-10769-6_31
15. Durán, F., Rocha, C., Salaün, G.: Symbolic specification and verification of data-aware BPMN processes using rewriting modulo SMT. In: International Workshop on Rewriting Logic and its Applications. LNCS, vol. 11152, pp. 76–97. Springer, Cham (2018). https://doi.org/10.1007/978-3-319-99840-4_5
16. Dutertre, B.: Yices 2.2. In: International Conference on Computer Aided Verification. LNCS, vol. 8559, pp. 737–744. Springer, Cham (2014). https://doi.org/10.1007/978-3-319-08867-9_49

17. Gao, S., Kong, S., Clarke, E.M.: dReal: An SMT solver for nonlinear theories over the reals. In: International Conference on Automated Deduction. LNCS, vol. 7898, pp. 208–214. Springer, Berlin (2013). https://doi.org/10.1007/978-3-642-38574-2_14
18. Lee, J., Bae, K., Ölveczky, P.C.: An extension of HybridSynchAADL and its application to collaborating autonomous UAVs. In: International Symposium on Leveraging Applications of Formal Methods, Verification and Validation. Adaptation and Learning. LNCS, vol. 13703, pp. 47–64. Springer, Cham (2022). https://doi.org/10.1007/978-3-031-19759-8_4
19. Lee, J., Bae, K., Ölveczky, P.C., Kim, S., Kang, M.: Modeling and formal analysis of virtually synchronous cyber-physical systems in AADL. Int. J. Softw. Tools Technol. Transfer **24**(6), 911–948 (2022). https://doi.org/10.1007/s10009-022-00665-z
20. Lee, J., Kim, S., Bae, K., Ölveczky, P.C.: HybridSynchAADL: Modeling and formal analysis of virtually synchronous CPSs in AADL. In: International Conference on Computer Aided Verification. LNCS, vol. 12759, pp. 491–504. Springer, Cham (2021). https://doi.org/10.1007/978-3-030-81685-8_23
21. Lee, J., Kim, S., Bae, K.: Bounded model checking of PLC ST programs using rewriting modulo SMT. In: International Workshop on Formal Techniques for Safety-Critical Systems, pp. 56–67. ACM (2022).https://doi.org/10.1145/3563822.3568016
22. Meseguer, J.: Generalized rewrite theories, coherence completion, and symbolic methods. J. Logical Algebraic Methods Program. **110**, 100483 (2020). https://doi.org/10.1016/j.jlamp.2019.100483
23. Microsoft Corporation: Strategies. https://microsoft.github.io/z3guide. Accessed 25 Mar 2022
24. Nigam, V., Talcott, C.: Automating safety proofs about cyber-physical systems using rewriting modulo SMT. In: International Workshop on Rewriting Logic and its Applications, LNCS, vol. 13252, pp. 212–229. Springer, Cham (2022). https://doi.org/10.1007/978-3-031-12441-9_11
25. Rocha, C., Meseguer, J., Muñoz, C.: Rewriting modulo SMT and open system analysis. J. Logical Algebraic Methods Program. **86**(1), 269–297 (2017). https://doi.org/10.1016/j.jlamp.2016.10.001
26. Rubio, R.: Maude as a library: An efficient all-purpose programming interface. In: International Workshop on Rewriting Logic and its Applications. LNCS, vol. 13252, pp. 274–294. Springer, Cham (2022). https://doi.org/10.1007/978-3-031-12441-9_14
27. Urquiza, A.A., AlTurki, M.A., Kanovich, M., Kirigin, T.B., Nigam, V., Scedrov, A., Talcott, C.: Resource-bounded intruders in denial of service attacks. In: Computer Security Foundations Symposium, pp. 382–38214. IEEE (2019). https://doi.org/10.1109/CSF.2019.00033
28. Wang, Q., Datta, P., Yang, W., Liu, S., Bates, A., Gunter, C.A.: Charting the attack surface of trigger-action IoT platforms. In: ACM Conference on Computer and Communications Security, pp. 1439–1453. ACM (2019). https://doi.org/10.1145/3319535.3345662
29. Whitters, G., Nigam, V., Talcott, C.: Incremental rewriting modulo SMT. In: International Conference on Automated Deduction. LNCS, vol. 14132, pp. 560–576. Springer, Cham (2023). https://doi.org/10.1007/978-3-031-38499-8_32
30. Yu, G., Bae, K.: Maude-SE: a tight integration of Maude and SMT solvers. In: International Workshop on Rewriting Logic and its Applications (2020)

Education Papers

Teaching an Advanced Maude-Based Formal Methods Course in Oslo

Peter Csaba Ölveczky[✉]

Department of Informatics, University of Oslo, Oslo, Norway
`peterol@ifinie.no`

Abstract. I have previously described an introductory Maude-based formal methods course in Oslo. In this paper, I describe a follow-up "advanced" rewriting-logic-based formal methods course. It consists of three assignments, a few theoretical topics, and a number of topics for student presentations that should illustrate the wide range of domains in which formal methods, including rewriting logic, have been successfully applied. I describe the course content, and evaluate the different topics based on my own impressions, exams, and student feedback.

1 Introduction

I have in two invited papers [22,27] argued why rewriting logic [17] and its modeling language and analysis tool Maude [12] should be an excellent choice for teaching formal methods, according to criteria in, e.g., [1,9,28]. The reasons include: (i) modeling is fun object-oriented and functional programming, which students tend to like; (ii) students can easily model and analyze key distributed algorithms and protocols taught in other courses (e.g., databases, security, distributed systems, operating systems, etc.) so that such a formal methods course can be well integrated with other courses in a CS curriculum; (iii) emphasis on *automated* analyses; and (iv) *one* formal method covers a lot of ground.

I have taken my own medicine and have been teaching an introductory Maude-based formal methods course at the University of Oslo for twenty years, and have even written a textbook, *Designing Reliable Distributed Systems: A Formal Methods Approach Based on Executable Modeling in Maude* [26], for that course. The topics treated in such a first formal methods course are almost given: specifying data types, termination, confluence, and equational logic, modeling distributed systems in an object-oriented way on a number of classic distributed systems applications, proving invariants, and introduce temporal logic and temporal logic model checking (see Sect. 2).

Having covered most of the mandatory basics in the introductory course, we have much more freedom in deciding what to teach in a follow-up "advanced" course for M.Sc and Ph.D. students; in addition, some third-year B.Sc students also take the course, and tend to do well. In this paper I explain what I teach in such an advanced 10-credit Maude-based course.

As a course intended to prepare students for their MSc. studies, one goal is to make students read, understand, and present scientific papers, maybe for the first time. Another goal is to present fundamental formal methods concepts and methods, and yet another goal is to show how formal methods can be used on state-of-the-art problems and showcasing impressive applications of formal methods. Furthermore, while the course is based on Maude to ground the new stuff in something the students (should) know well, the intention is to teach *general* formal methods topics instead of very rewriting-logic-specific ones.

In this paper, I give a brief background on the content of the basic Maude course in Sect. 2 to present what the students already (should) know. In Sect. 3 I present the topics taught in this advanced course. I try to explain *why* I teach the topic, *what* we study in each topic, and then give a brief *evaluation* of how well it works, based on own impressions, exams, and feedback from students. Finally, I give some overall impressions and student feedback in Sect. 4.

The goal of this paper is to: (i) present and motivate some topics which might pique the interest of other on what to teach; (ii) share my experiences; and (iii) first and foremost, I hope that this will inspire others, e.g., at WRLA 2024, to discuss what they are teaching, and give me ideas of what I should teach in this course the next time. I believe that many of us are using Maude in our teaching, and sharing our experiences and ideas would greatly benefit our community.

2 Context

I briefly present the content of the basic Maude course at the University of Oslo for two reasons: To give an impression of the students' (presumed) background and to show the content that therefore cannot be part of the advanced course. I then present some practical information about the course and its students.

Introductory Maude-based Formal Methods Course in Oslo. The first part of the book/course[1] covers equational specification in Maude and the algebraic specification classics: term rewriting, termination (undecidability*), progress functions, the theory of simplification orders*, and the basic path orders), confluence, equational logic and inductive theorems, and basic algebra/model theory*.

The second part covers modeling distributed systems in rewriting logic and analyzing them in Maude: rewriting logic, object-oriented specification in Maude, and linear temporal logic (LTL) and LTL model checking in Maude. The systems modeled include transport protocols ("TCP," alternating bit, and sliding window), distributed systems classics (two-phase commit and well-known distributed mutual exclusion and leader election protocols), and security protocols (we model and break NSPK with a simple Maude search).

Setting. This is a 10 ECTS credits course with one lecture—lasting between 90 and 135 min—per week for seven weeks, followed by student presentations. The course is typically taken by around eight or nine reasonably talented students.

---

[1] Parts of the book not taught to second-year students are marked with '*'.

Surprisingly many of the students have not taken the basic Maude course, but tend to do well anyways, since the course starts with a lecture summarizing the modeling and analysis of distributed systems in Maude. One limiting factor is that, according to a recent survey, MSc students in this program spend on average 23.6 h/week on their studies.

3 Course Content

The course consists of the following three "components":

1. Fundamental formal methods topics.
2. Three mandatory assignments.
3. A number of topics to be studied and presented by students.

The following sections outline the content and curriculum in each category.

3.1 Theoretical Topics

Real-Time Systems (1 lecture). Real-time systems can be modeled in rewriting logic as real-time rewrite theories [17], whose modeling and analysis are supported by the Real-Time Maude tool [21,23,24]. The required reading is my WRLA 2014 invited paper *Real-Time Maude and its Applications* [21].

We cover the basics, with small examples like breakable watches, populations, and round trip time protocols, and explain how temporal logics can be extended with time. We discuss the wide range of applications of Real-Time Maude to distributed real-time systems, including two avionics applications, and its role as a semantics framework and formal analysis backend to industrial modeling languages such as subsets of Ptolemy II and the avionics standard AADL [20].

Evaluation. My students have not commented on this topic; they rather tend to have opinions about applications. Real-time systems are fundamental, and most modeling formalisms have extensions to the timed case. Somewhat surprisingly, students tend to do better on other topics in the exam.

Probabilistic Systems (1 lecture). Probabilistic systems are key to reason about probabilistic events (needed, e.g., to certify aircrafts) and for performance estimation of distributed systems, where key "environment" parameters, such as message delays, can be assumed to follow certain probability distributions.

In this topic, whose required reading is the papers *PMaude: Rewrite-based Specification Language for Probabilistic Object Systems* [2] and *PVeStA: A Parallel Statistical Model Checking and Quantitative Analysis Tool* [3], we cover the very basics: (Discrete time) Markov chains and Markov decision processes with small examples. We discuss why finding probabilities amounts to finding the expected value of an expression over a behavior (the probability that Federer wins a service game can be reduced to asking how many service games Fed is expected to win if he plays hundred games). We discuss probabilistic model checking and statistical model checking and their differences. We exemplify probabilistic rewrite theories on small examples like Federer's service game, England in penalty shoot-outs, and human aging, and discuss how such probabilistic rewrite theories can be simulated in Maude. We then look into probabilistic and stochastic temporal logic, statistical model checking, and the Actor PMaude approach to almost-surely obtaining purely probabilistic (real-time) systems (needed for statistical model checking) by sampling message delays from continuous distributions.

Evaluation. This topics works well and covers a lot of conceptual ground without going into much detail. To my chagrin the slides look more interesting than the ones for real-time systems, and, surprisingly, students tend to do better on this topic than on real-time systems.

Metaprogramming (2 lectures). Metaprogramming means that programs can be seen as data that can be manipulated by programs (that ...). In our setting it means (meta-)representing Maude modules/specifications as Maude terms (of sort `Module`) that can be manipulated by Maude functions on such terms. The curriculum on this topic is the appropriate chapter of the Maude manual [11]. The first lecture covers meta-representing modules and defining simple functions (union of two modules, add/remove a rule to/from a module, check whether a module is lpo-terminating, and so on) on such modules. The second lecture defines simple execution strategies using `META-LEVEL` operators such as `metaXapply`, `metaReduce`, and `metaSearch`, which should culminate in Assignment 2, where the students implement randomized simulation in Maude.

Evaluation. I have been very pleasantly surprised by how quickly and easily students learn metaprogramming in Maude.

3.2 Mandatory Assignments

Assignment 1: Modeling and Analyzing a Broadcast Protocol. In their first assignment, students model and analyze in Maude the *reliable broadcast protocol* (RBP), but only for *static* networks, developed in 1995 by the group of a leading researcher in distributed systems, J. J. Garcia-Luna at UCSC, and published in the paper *Reliable Broadcasting in Dynamic Networks* [15].

Motivation. The goals of this assignment are to: (i) refresh or introduce Maude modeling and analysis of distributed systems, and (ii) illustrate the use and benefits of formal methods for distributed systems.

The protocol is very simple in the static network setting, and is therefore a good refresher (or introduction) to Maude. More importantly, it illustrates that a protocol description published in a good conference by a star researcher, containing both precise-looking pseudo-code and a proof of correctness is (a) highly ambiguous and full of missing assumptions, and (b) easily can be found incorrect with simple Maude analysis.

Task. Model the RBP protocol (in the simplified static setting) in Maude, using Maude's object-based specification methods; discuss whether the protocol description in the paper contains important implicit assumptions and/or ambiguities; define some suitable initial states and analyze RBP using Maude, and try to figure out whether or not the protocol is correct. Finally, the students should discuss and compare the two description and validation methods, namely, formal specification and automated model checking analysis versus informal descriptions with simulations and hand proofs of correctness, and list some advantages and disadvantages of each method.

Evaluation. This is an excellent task to start the course: it is easy to specify RBP in Maude; there are many ambiguities and missing assumptions in the paper; and a search for deadlocks in a three-node setting finds that the protocol is incorrect, so that the "hand-proven" correctness theorem does not hold. It illustrates on a very simple example the benefits of formal methods.

The main disadvantage is that the protocol is very old and almost too simple. It boils down to balancing between finding ambiguities and errors in proven protocols of more modern and complex systems versus how much effort should go into the task. One suggestion for a smallish but still more complex and modern protocol with the same issues could be the *P-Store* distributed transaction system design [30], whose errors, ambiguities, and missing assumptions I point out in [25]. However, P-Store requires modeling *atomic broadcast*.

I will definitely keep this assignment for the next time. It is a small and illuminating introduction to Maude and the usefulness of formal methods.

Assignment 2: Metaprogramming in Maude. In this assignment the students define new generic analysis methods for Maude. Lately, this analysis method has been *randomized simulation*.

Motivation. The goal is to get experience with metaprogramming, and to show that useful generic analysis methods easily can be implemented in this way. It also prepares them for statistical model checking, which is covered later.

Task. Define a metalevel function for single randomized simulations, and then a function for performing *many* randomized simulations. Test the commands on small examples such as the football game [26] or the "whiteboard game."

Evaluation. This task has a short and simple solution, and provides in about half a page of Maude code a very useful simulation method not directly supported by Maude. Students tend do to quite well in metaprogramming, although surprisingly few solve it using `metaSearch`, which seems to be the easiest and most elegant way. I previously asked the students to program "iterative bounded depth-first search," but that was significantly more complicated; furthermore, it seemed less well motivated since it performed worse than Maude search on the NSPK protocol. I am happy with the current assignment.

Assignment 3: Maude Semantics for an Imperative Programming Language. In this assignment the students define a Maude semantics/interpreter for a small imperative toy language.

Motivation. Introduce Maude semantics of imperative programming languages, how a formal semantics is important to clarify the meaning of programming languages, and also how a high-level formal semantics allows us to quickly experiment with different programming language designs and/or variations of the meaning of different constructs. It should also prepare the ground for the topics of programming language semantics in the $\mathbb{K}$ framework, and its use to define and analyze C programs and Ethereum programs/contracts. In this assignment, the students get acquainted with concepts such as cells and continuations.

Task. Define the Maude semantics of, and thereby an interpreter for, an imperative toy programming language, called Borneo. A Borneo program is a set of method definitions, with one method called `main`. A method may have input parameters and local variables, and may return a value. The statements are: assignments of an integer expression to a variable, call a method, if tests, while loops, and sequencing. Expressions are simple integer and Boolean expressions.

The students should test their interpreter on Borneo imperative programs with recursive functions for, e.g., primality testing, computing the factorial or a Fibonacci number, and finding the greatest common divisor.

Evaluation. Balancing what language features to include with workload and elegance is the main tradeoff in such a task. On the one hand, I *really* wanted to include arrays, to introduce heaps, call by reference, memory management, aliasing, and so on. However, after defining the semantics I thought it would be slightly too much for a small assignment. It would also have been great to introduce threads. Multithreaded programs *without* methods can be given a very short and elegant semantics in Maude, which I already did for an exam in the second-year course. With threads one can easily define multithreaded programs for, e.g., mutual exclusion, and can use Maude model checking to great effect.

Apart from not introducing threads and/or arrays, another weakness with this assignment is the lack of program verification. We do not introduce symbolic reasoning in this course, which limits what we can do with our Maude semantics, apart from running the programs on concrete inputs.

The assignment works well, but I should work more on finding the insight-vs-effort/elegance sweet spot in terms of which imperative programming language features to include.

3.3 Topics for Student Presentations

The purpose of the student presentations is to illustrate the wide range of applications of formal methods in general and of rewriting logic in particular.

The challenge here is to find a balance between the difficulty of the papers, the level of detail (some concrete detail, like a few Maude rules are great to keep it concrete), appeal of application domain, elegance, impact of results, and so on. This part is obviously where the choice of possible topics is by far the greatest, and where I especially would appreciate good suggestions!

Each topic is presented by one student in a 45-minute presentation.

Rewriting Logic Models of Cells and Biological Networks. Formal methods are increasingly used in systems biology, and SRI International had for many years a project on using rewriting logic to model biological entities and processes. In essence, mammalian cells and proteins are modeled in rewriting logic at a very pleasant level of abstraction. Rewrite rules then define how various proteins move in and out of the various parts of a cell, and how they aggregate and transform.

The challenge is to find papers which present the rewriting logic models of a cell as terms and the biological reactions as rewrite rules. For this requirement of a more detailed view of the rewriting model, I have found that the early paper *Pathway Logic: Executable Models of Biological Networks* [13] is an excellent source which makes it easy to understand these models. This paper contains signatures for representing the biological entities and fairly simple and understandable rewrite rules for modeling biological processes. The paper includes understandable examples of Maude analysis, including LTL model checking (although it is hard to understand why search was not sufficient). I also recommend looking over Carolyn Talcott's slides from ISR 2021 and José Meseguer's slides on *Bio-Pathway Logic*.

Evaluation. In general this topic works well; with some effort one can easily understand the Maude formalizations. The cited paper is very nice for this purpose. Students tend to do very well on this topic in the exam.

Whereas there is a large amount of interesting topics to choose from for the student presentations, I foresee that this topic will remain also the next time.

Modeling and Analyzing Protein Aggregation in the Brain. Continuing with systems biology, the course also has a topic on modeling aggregation (proteins grouping together) and dissolution of proteins in the brain, where large aggregates (called *Lewy bodies*) are thought contribute to the onset of Parkinson's Disease. In the paper *Using Probabilistic Strategies to Formalize and Compare α-Synuclein Aggregation and Propagation under Different Scenarios* [7], by myself, a then-PhD student of mine, and a researcher in medicine, we model how proteins aggregate and dissolve in the brain, how they move in the brain, how they cause the death of single neurons, and so on. We then extend the model with different probabilities of applying the different rules, based on a person's age, predisposition to Parkinson's, and intake of an experimental medicine, and perform randomized simulations to simulate the amount of Lewy bodies in different brain regions over time in the three different scenarios.

Evaluation. This topic works well. The students have already had lectures on both timed and probabilistic extensions of rewriting logic. Furthermore, the rules are small and easy to understand, and so are the "weights" assigned to the different rules in the different scenarios.

The reason for being skeptical is whether such modeling of *single* proteins and neurons, and performing randomized simulations provide useful analyses of brains. Furthermore, even with one coauthor being an accomplished researcher in medicine at KU Leuven, some might question the credentials of the authors.

The paper works well for teaching, should be easily understandable, has enough details for good examinations, and the students seem to be reasonable happy with it. Nevertheless, because of the doubt about whether this represents the best methodology in the field, and because having two systems biology topics in the course might be too much, this topic might be replaced.

Formal Semantics of C in $\mathbb{K}$. Grigore Roşu and his collaborators have for a number of years done very impressive work on defining the formal semantics of programming languages in rewriting logic through their $\mathbb{K}$ framework. Among the nice early papers, including survey papers and a strong paper on the semantics of Java, the paper I thought fit best for this course is the POPL 2012 paper

An executable formal semantics of C with applications [14]. This paper gives a brief background on language semantics definition in 𝕂, shows some examples of such 𝕂 rules, explains the various ways in which C is "undefined." The paper also gives nice and easily understandable concrete examples of the consequences of such undefinedness. I did not find a paper on 𝕂 that I thought would fit my students, but fortunately this paper provides a very useful brief introduction and intuition to the 𝕂 way of defining programming language semantics.

Evaluation. This paper works well. It is easy to read and understand and gives a short and useful introduction to 𝕂 anno 2012. Although this introduction is short, fortunately the students defined the Maude semantics of Borneo earlier in the course (see above), and could therefore understand key 𝕂 concepts such as cells and continuations, etc. The only very minor disadvantage from the point of view of this course is the lack of concrete rewriting logic code/specification.

Formal Semantics of the Ethereum Virtual Machine in 𝕂. Whereas the C semantics paper [14] represented early use of 𝕂, that 𝕂 framework now comes with a compelling vision (and accompanying tool support) of how defining the formal semantics of a programming language (which must be done anyways) gives you 𝕂-semantics-based tools such as interpreter, debuggers, simulators, model checkers, theorem provers, etc., *for free*.

One of the commercial application areas where this framework is used is electronic contracts on the (Ethereum) blockchain. The selected paper in this course is *KEVM: A Complete Formal Semantics of the Ethereum Virtual Machine* [16]. In addition, the reading list for this topic also contains the following short papers about the 𝕂 vision: *The K Vision for the Future of Programming Language Design and Analysis* [10] and Roşu's invited FSCD paper *Formal Design, Implementation and Verification of Blockchain Languages*.

Evaluation. This topic works well, and does not require detailed knowledge of 𝕂, but the vision is inspiring and should be easy to understand. The student presenting this topic last year has chosen to formalize online trading platforms in Maude for his MSc. The problem might be having two topics on 𝕂 and programming language semantics. If that must be reduced, I would keep this topic instead of the C semantics (because of the sexier topic and "the vision thing").

Cloud-based Transaction Systems in Maude. Maude and its accompanying tools have been used to analyze and compare design options for a number of industrial and academic transaction systems in the context of University of Illinois Center for Assured Cloud Computing. Industrial systems included Google's Megastore, Apache Cassandra, and Apache Zookeeper, and sophisticated academic systems included UC Berkeley's RAMP transaction systems and the P-Store design. The many papers on the topic are summarized in the paper *Survivability: Design, Formal Modeling, and Validation of Cloud Storage Systems using Maude* [8], which was one of the required papers on this topic.

The reasons for selecting this topic includes: (i) use of Maude on large industrial systems; (ii) illustrates how Maude can compare different design options very early, to quickly discard proposed designs that do not seem improve the system's functionality or performance; (iii) students want computer science applications; and (iv) it shows how statistical model checking, treated earlier in the course, can be used for early model-based performance estimation of different design choices, and demonstrates the predictive power of these model-based performance comparisons, which correspond very well with the performance of their distributed implementations on industrial workloads.

Maybe the main reason for including this topic is that it gives me the excuse to put the very nice paper *How Amazon Web Services Uses Formal Methods* [19] on the list of required reading. This paper was written by engineers at Amazon Web Services, and makes an excellent case for formal methods during the design of sophisticated distributed systems.

Evaluation. Good applications summarized in one paper is nice, but in such a course some concrete Maude detail/code would be welcome. But the results are nice, and the Amazon paper is a gem, so I will probably keep this topic.

Breaking Internet Explorer Using Maude. One of the first "killer apps" of Maude was finding previously unknown address bar and status bar attacks in the Internet Explorer web browser. These vulnerabilities could cause you to see the wrong url in the browser's address bar, which has obvious security implications. Ralf Sasse, then a PhD student in the Maude group at UIUC, together with researchers at Microsoft, modeled in Maude how mouse moves over, and clicks on, elements of the web page affected various components of the browser. They then used Maude search to find unfortunate situations, like an address bar with a different url than the web site displayed in the browser. These findings are described in the required reading on this topic, *A Systematic Approach to Uncover Security Flaws in GUI Logic* [18].

Evaluation. Even today this work remains one of the most significant applications of Maude, which obviously means that it is natural to include it in such a course. The paper is very nice, but for teaching purposes it may suffer from being a too substantial contribution that may not fit well with the desire to understand easy-to-grasp smaller pieces of concrete code.

Breaking EMV Card Payment Using the Tamarin Prover. We go from one of the most remarkable applications of Maude to one of the most remarkable applications of formal methods: the work by David Basin, Ralf Sasse, and Jorge Toro on using the Tamarin prover to break the Europay/MasterCard/VISA payment system. In their paper *The EMV Standard: Break, Fix, Verify* [6], they demonstrate that you can bypass typing your PIN code when paying for expensive items, and that you can actually "pay" without being charged. A Norwegian newspaper even made a video of their work (https://www.vgtv.no/video/205022/ny-svindelmetode-slik-kan-de-kopiere-kortet-ditt-i-koeen).

I justified including this work into the course because the Tamarin prover is based on multiset rewriting, and uses some components from Maude. This topic therefore also included a quick understanding of the Tamarin prover; in particular of its impressive array of applications, such as finding faults in the 5G standard, via the paper *Tamarin: Verification of Large-Scale, Real-World Cryptographic Protocols* [5].

Evaluation. Again, the amazing results mean that the (counter)examples are large and difficult to grasp for someone not well versed in how payment protocols work. Furthermore, reading papers in a different notation might be too much for our students. I am therefore not sure whether this remarkable formal methods application will be on the curriculum next time.

Monitoring. Monitoring, or runtime verification, is a recent popular scalable formal method. I thought it would be interesting to introduce students to a new kind of formal method, and selected the following overview paper of monitoring, *Introduction to runtime verification* [4]. To justify including this topic into a rewriting-logic-based course, I also included Rosu and Havelund's paper *Rewriting-based techniques for runtime verification* [29].

Evaluation. The overview paper was nice. The technical paper focused on how reading an input affected the temporal logic formula being monitored. This somehow seemed too little, and maybe not sufficiently interesting, for a presentation, so this topic will probably not be given next time.

4 My Impressions and Student Feedback

My Impressions. One of my worries is that I focus too much on applications and not enough on new formal methods concepts. What other *general* formal methods concepts should I introduce? *Symbolic analysis* is important in computer science in general, is well supported by Maude through both narrowing and rewriting modulo SMT, and should be a good candidate if I can find small but elegant and compelling applications. This would also make it easier to understand the Tamarin tool. What else? Theorem proving of various kinds?

Another kind of topic that might be interesting to explore is to see whether Maude's support for interacting with the external world through sockets and operating systems calls can be exploited in some fun project. Suggestions and experiences are most welcome!

Student Feedback. Since the course has only been given twice, each time to 8–9 students, I do not have significant formal feedback. Based on the little feedback, and discussions with the students, it seems that my worries about the course being too easy and having too low workload do not hold. In general the students like the focus on applications, and they do not miss more "theory." I thought that including all this biology/neuroscience stuff would be popular, but students prefer "less biology, more computer science." The most popular topic(s) seem to have been the formal semantics of C and KEVM parts; a number of students even read up on blockchains when studying the KEVM paper. Students thought that the papers of the most impressive applications, breaking EMV payments with the Tamarin prover and breaking Internet Explorer with Maude, were "more difficult to understand." This is likely because these papers did not present a few small and simple rewrite rules to ground their excellent stories.

Students liked studying and presenting research papers, which they have not done in other courses in Oslo. They loved the first and third assignments, but some thought that the second assignment was too small (and somewhat tricky).

5 Concluding Remarks

I have presented the contents and impressions and anecdotal student feedback on a generally well-received "advanced" course in formal methods based on Maude at the University of Oslo. I hope that this paper can inspire others teaching Maude, and that it will generate interesting discussions at WRLA 2024, where we can share our experiences in teaching Maude-based formal methods courses.

Acknowledgments. I thank the anonymous WRLA 2024 reviewers for their insightful comments on a previous version of this paper, and I gratefully acknowledge financial support by the NATO Science for Peace and Security Programme through grant number G6133 (project SymSafe).

References

1. Aceto, L., Ingólfsdóttir, A., Larsen, K.G., Srba, J.: Teaching concurrency: Theory in practice. In: Proceedings of Teaching Formal Methods 2009. LNCS, vol. 5846. Springer, Berlin (2009)
2. Agha, G.A., Meseguer, J., Sen, K.: PMaude: Rewrite-based specification language for probabilistic object systems. Electr. Notes Theor. Comput. Sci. **153**(2) (2006)
3. AlTurki, M., Meseguer, J.: PVeStA: A parallel statistical model checking and quantitative analysis tool. In: CALCO'11, LNCS, vol. 6859. Springer, Berlin (2011)
4. Bartocci, E., Falcone, Y., Francalanza, A., Reger, G.: Introduction to runtime verification. In: Lectures on Runtime Verification: Introductory and Advanced Topics, LNCS, vol. 10457, pp. 1–33. Springer, Berlin (2018)
5. Basin, D.A., Cremers, C., Dreier, J., Sasse, R.: Tamarin: verification of large-scale, real-world, cryptographic protocols. IEEE Secur. Priv. **20**(3), 24–32 (2022)
6. Basin, D.A., Sasse, R., Toro-Pozo, J.: The EMV standard: break, fix, verify. In: 42nd IEEE Symposium on Security and Privacy, SP 2021. IEEE (2021)
7. Bentea, L., Ölveczky, P.C., Bentea, E.: Using probabilistic strategies to formalize and compare α-synuclein aggregation and propagation under different scenarios. In: Proceedings of Computational Methods in Systems Biology (CMSB'13). LNCS, vol. 8130. Springer, Berlin (2013)
8. Bobba, R., Grov, J., Gupta, I., Liu, S., Meseguer, J., Ölveczky, P.C., Skeirik, S.: Survivability: design, formal modeling, and validation of cloud storage systems using Maude. In: Assured Cloud Computing, Chap. 2, pp. 10–48. Wiley-IEEE Computer Society Press (2018)
9. Cerone, A., et al.: Rooting formal methods within higher education curricula for computer science and software engineering: A white paper. In: Proceedings of FMfun 2019, First International Workshop on Formal Methods—Fun for Everybody. Communications in Computer and Information Science (CCIS), vol. 1301. Springer, Berlin (2020)
10. Chen, X., Rosu, G.: The K vision for the future of programming language design and analysis. In: Formal Methods in Outer Space: Essays Dedicated to Klaus Havelund on the Occasion of His 65th Birthday. Lecture Notes in Computer Science, vol. 13065, pp. 3–9. Springer, Berlin (2021)

11. Clavel, M., Durán, F., Eker, S., Escobar, S., Lincoln, P., Martí-Oliet, N., Meseguer, J., Rubio, R., Talcott, C.: Maude Manual (Version 3.3.1) (2023). http://maude.cs.illinois.edu
12. Clavel, M., Durán, F., Eker, S., Lincoln, P., Martí-Oliet, N., Meseguer, J., Talcott, C.: All About Maude, LNCS, vol. 4350. Springer, Berlin (2007)
13. Eker, S., Knapp, M., Laderoute, K., Lincoln, P., Talcott, C.L.: Pathway logic: Executable models of biological networks. In: Proceedings of WRLA 2002. Electronic Notes in Theoretical Computer Science, vol. 71, pp. 144–161. Elsevier (2002)
14. Ellison, C., Rosu, G.: An executable formal semantics of C with applications. In: Proceedings of the 39th ACM SIGPLAN-SIGACT Symposium on Principles of Programming Languages, POPL 2012, pp. 533–544. ACM (2012)
15. Garcia-Luna-Aceves, J., Zhang, Y.: Reliable broadcasting in dynamic networks. In: Proceedings of IEEE ICC. IEEE (1996)
16. Hildenbrandt, E., Saxena, M., Rodrigues, N., Zhu, X., Daian, P., Guth, D., Moore, B.M., Park, D., Zhang, Y., Stefanescu, A., Rosu, G.: KEVM: A complete formal semantics of the ethereum virtual machine. In: 31st IEEE Computer Security Foundations Symposium, CSF 2018, pp. 204–217. IEEE Computer Society (2018)
17. Meseguer, J.: Conditional rewriting logic as a unified model of concurrency. Theor. Comput. Sci. **96**, 73–155 (1992)
18. Meseguer, J., Sasse, R., Wang, H.J., Wang, Y.: A systematic approach to uncover security flaws in GUI logic. In: 2007 IEEE Symposium on Security and Privacy (S&P 2007). IEEE Computer Society (2007)
19. Newcombe, C., Rath, T., Zhang, F., Munteanu, B., Brooker, M., Deardeuff, M.: How Amazon Web Services uses formal methods. Commun. ACM **58**(4), 66–73 (2015)
20. Ölveczky, P.C.: Semantics, simulation, and formal analysis of modeling languages for embedded systems in Real-Time Maude. In: Agha, G., Danvy, O., Meseguer, J. (eds.) Talcott Festschrift, Lecture Notes in Computer Science, vol. 7000, pp. 368–402. Springer, Berlin (2011)
21. Ölveczky, P.C.: Real-Time Maude and its applications. In: WRLA 2014. LNCS, vol. 8663. Springer, Berlin (2014)
22. Ölveczky, P.C.: Teaching formal methods for fun using Maude. In: Proceedings of FMfun 2019, First International Workshop on Formal Methods—Fun for Everybody. Communications in Computer and Information Science (CCIS), vol. 1301, pp. 58–91. Springer, Berlin (2020)
23. Ölveczky, P.C., Meseguer, J.: Semantics and pragmatics of Real-Time Maude. Higher-Order Symbolic Comput. **20**(1–2), 161–196 (2007)
24. Ölveczky, P.C., Meseguer, J.: The Real-Time Maude tool. In: Proceedings of TACAS'08. LNCS, vol. 4963. Springer, Berlin (2008)
25. Ölveczky, P.C.: Formalizing and validating the P-Store replicated data store in Maude. In: Proceedings of WADT'16. LNCS, vol. 10644. Springer, Berlin (2016)
26. Ölveczky, P.C.: Designing Reliable Distributed Systems: A Formal Methods Approach Based on Executable Modeling in Maude. Undergraduate Topics in Computer Science. Springer, Berlin (2017)
27. Ölveczky, P.C.: Teaching formal methods to undergraduate students using Maude. In: Rewriting Logic and Its Applications (WRLA@ETAPS 2022). LNCS, vol. 13252. Springer, Berlin (2022)
28. Roggenbach, M., Cerone, A., Schlingloff, B., Schneider, G., Shaikh, S.A.: Formal Methods for Software Engineering: Languages, Methods, Application Domains. Texts in Theoretical Computer Science. An EATCS Series. Springer, Berlin (2022)

29. Rosu, G., Havelund, K.: Rewriting-based techniques for runtime verification. Autom. Softw. Eng. **12**(2), 151–197 (2005)
30. Schiper, N., Sutra, P., Pedone, F.: P-Store: Genuine partial replication in wide area networks. In: 29th IEEE Symposium on Reliable Distributed Systems (SRDS 2010), pp. 214–224. IEEE Computer Society (2010)

Author Index

MISC
Ölveczky, Peter Csaba   124, 195

B
Bae, Kyungmin   3, 179
Ban Kirigin, Tajana   22

C
Comer, Jesse   22

E
Escobar, Santiago   3

D
Do, Canh Minh   84

K
Kanovich, Max   22

L
Lienhardt, Michael   165
López-Rueda, Raúl   3

M
Martí-Oliet, Narciso   145
Meseguer, José   3, 62

O
Ogata, Kazuhiro   45, 84
Olarte, Carlos   104, 124

P
Pita, Isabel   145

R
Ramírez, Carlos   104
Rocha, Camilo   104
Rubio, Rubén   145

S
Sapiña, Julia   3
Scedrov, Andre   22

T
Talcott, Carolyn   22
Tran, Duong Dinh   45

V
Valencia, Frank   104
Verdejo, Alberto   145

Y
Yu, Geunyeol   179

SPRINGER NATURE

GPSR Compliance

The European Union's (EU) General Product Safety Regulation (GPSR) is a set of rules that requires consumer products to be safe and our obligations to ensure this.

If you have any concerns about our products, you can contact us on ProductSafety@springernature.com

In case Publisher is established outside the EU, the EU authorized representative is:

Springer Nature Customer Service Center GmbH
Europaplatz 3
69115 Heidelberg, Germany

The manufacturer's authorised representative in the EU is Springer Nature Customer Service Centre GmbH, Europaplatz 3, 69115 Heidelberg, Germany. If you have any concerns regarding our products, please contact ProductSafety@springernature.com

Printed and bound by CPI Group (UK) Ltd, Croydon, CR0 4YY

25/03/2026

02078187-0008